Open Minded

Open Minded: Searching for Truth about the Unconscious Mind

Ben R. Newell and David R. Shanks

The MIT Press
Cambridge, Massachusetts
London, England

The MIT Press would like to thank the anonymous peer reviewers who provided comments on drafts of this book. The generous work of academic experts is essential for establishing the authority and quality of our publications. We acknowledge with gratitude the contributions of these otherwise uncredited readers.

This book was set in Stone Serif and Stone Sans by Westchester Publishing Services. Printed and bound in the United States of America.

Library of Congress Cataloging-in-Publication Data

Names: Newell, Benjamin R., 1972– author. | Shanks, David R.
Title: Open minded : searching for truth about the unconscious mind / Ben R. Newell and David R. Shanks.
Description: Cambridge, Massachusetts : The MIT Press, [2023] | Includes bibliographical references and index.
Identifiers: LCCN 2022038725 (print) | LCCN 2022038726 (ebook) | ISBN 9780262546195 (paperback) | ISBN 9780262375368 (epub) | ISBN 9780262375375 (pdf)
Subjects: LCSH: Subconsciousness. | Thought and thinking. | Self-consciousness (Awareness)
Classification: LCC BF315 .N479 2023 (print) | LCC BF315 (ebook) | DDC 154.2—dc23/eng/20230316
LC record available at https://lccn.loc.gov/2022038725
LC ebook record available at https://lccn.loc.gov/2022038726

10 9 8 7 6 5 4 3 2 1

For my family, near and far, who are always on my mind—BRN
To Miranda, Will, and Ella—DRS

Contents

Preface

This book brings together two compelling ideas. The first is that behavioral scientists as well as public discourse place far more emphasis on the unconscious mind than is warranted by the evidence. The second is that science is going through a turbulent period of crisis and reappraisal, brought on by the realization that many of its methods and practices are indefensible. We describe the acute connection between these ideas and examine the ways in which poor scientific practices have supported pervasive but ultimately erroneous claims about the unconscious mind. Contrary to popular beliefs in the powers of the unconscious, we show that overwhelmingly people are conscious—in the sense of being aware—of the reasons underlying their behavior. This perspective provides a counter to claims gaining ever stronger traction in public debate about the ubiquity of unconscious influences on people's behavior. We hope that the book prompts a wider discussion in society about how we understand the mind.

* * *

The ideas in this book have grown out of many conversations over the twenty years that we have known each other. We started working together when Ben began his postdoctoral fellowship in David's lab at University College London (UCL) in 2001. We shared a fascination with the idea that behavior could be influenced by information that was entirely outside people's awareness. This idea has gained in popularity over the intervening two decades, despite increasingly shaky scientific foundations. Our frustration with this state of affairs is what led us to write this book. It is impossible to acknowledge all of the people who have influenced our thinking about these issues, so we will not attempt to name them for fear of missing some. Special thanks, however, are due to experts who generously gave up their time to read drafts of chapters: Jan De Houwer, Zoltan Dienes, Chris Donkin, Tom Hardwicke, Alice Mason, Simone Malejka, Craig McKenzie, Magda Osman, Aba Szollosi, Miguel Vadillo, and Eric-Jan Wagenmakers. Their comments were immensely helpful. Ben acknowledges the Australian Research Council for funding support, and the UNSW School of Psychology for providing a perfect working environment. David is grateful to the UK Economic and Social Research Council for grant support over many years, as well as to UCL for its extraordinary support for research in the behavioral and brain sciences. We also thank the editorial team at the MIT Press for their help throughout the publication process.

I The Search for the Unconscious Mind

1 Reclaiming the Science of the Mind

The mind is like an iceberg, it floats with one-seventh of its bulk above water.
—attributed to Sigmund Freud

In July 2017, iceberg A68 split from the Larsen C ice shelf on the eastern coast of the Antarctic peninsula. Despite its unassuming name, A68 covered an area almost four times the size of Greater London (approximately 6,000 square kilometers) and was one of the largest icebergs ever seen in the Antarctic. The calving of this giant 200-meter-thick berg, most of it hidden below the water, sent shockwaves through the scientific community and raised concerns that it might drift into shipping paths.[1]

The analogy of the human mind as an iceberg has been prominent in public discourse for a long time. At the turn of the previous century, Sigmund Freud popularized the idea that much of our mental life exists below the waterline of consciousness.[2] Our behavior, according to Freud, is driven by the desires and beliefs hidden in the depths of the unconscious. The real reasons for our behavior may be discovered only through years of psychoanalysis that allow these dangerous, murky, unconscious motives to bubble to the surface of consciousness. Contemporary perspectives tend to abandon the more colorful notions of repressed primitive urges but still emphasize the prominence of an adaptive, powerful, sophisticated unconscious mind that is essential for our survival in the world.[3]

Why is this idea so appealing, pervasive, and persistent? In this book, we argue that the enduring myth of the unconscious mind is a symptom of a much broader problem facing the social and behavioral sciences. Modern research on the science of mind and behavior has gone badly astray because of the culture in which scientists operate, which encourages tenuous research built on weak methods. Psychologists are rewarded for making eye-catching claims—we're unconsciously racist, ageist, ableist, and sexist, driven by external nudges rather than free will, for instance—that they hastily publish in highly prestigious scientific journals. Many of these claims turn out to be wrong.

What happened to A68? In the end, it broke up into fragments and melted, and perhaps this provides a more significant metaphor. As we discuss in this book, research conducted over the past few years, together with a much deeper understanding of the biases to which scientists are prone, suggest that the way we think and act can best be understood without invoking a powerful unconscious. The science of human behavior needs to be rebuilt from the ground up on firmer foundations.

* * *

The book is in two parts. Part I is a search for the unconscious mind. What is the evidence that our behavior is determined below the waterline and that a powerful unconscious is necessary for us to survive? In plumbing these depths, we carve out some key definitions for what would count, in our view, as an unconscious influence on behavior. Such definitions are crucial because many words have been devoted to the highly complex pursuit of characterizing our consciousness of various mental processes. Our focus throughout is on the core processes of decision making.

We take decision making to refer to the mental processing that leads to the selection of one among several actions (choices). This broad definition encompasses situations from judging where to run to on a field in order to catch a ball, to choosing jobs, houses, or even a spouse. In each situation, there is information to be perceived, processed, integrated, and acted on. The nature of that information is of course wildly different, but as we will see, in each case there are many similarities in how we make up our minds.

Construing decision making this way excludes examples such as neurons or brain networks making "decisions." Thus, the visual system's computation of low-level properties is not decision making on this definition. We view consciousness as a property of individuals and hence do not believe any useful purpose is served by asking whether the computation of motion in the brain, for instance, is or is not conscious. It is, in contrast, perfectly reasonable to ask whether an individual's judgment of motion is conscious.

Understanding this distinction between the products, or end results (a judgment of motion), and the processes by which that product was created (the firing of neurons in the motion processing part of the visual system, called area V5) is fundamentally important if we are to "invert the iceberg," We are never conscious of these low-level processes, but the products of these processes are available to consciousness and form the basis of our judgments and decisions.[4] This inverted view in which the vast majority, if not all, of thinking is above the waterline frees us to embark on a clear assessment of the evidence for unconscious thought and a radical reconceptualization of the basis of human decision making.

Not least, it allows us to delve more deeply into the idea that our decisions are controlled by two (or more) different systems that compete and conflict. One system is often characterized as automatic, reflexive, and operating largely outside our awareness, the other as deliberative, rational, and (cognitive) resource intensive. This dual-system view has had enormous influence in recent times, pervading not only psychology but also economics, medicine, business, and government.[5] We will see that this duality stems from the misguided iceberg metaphor, and when we invert the iceberg, evidence for dual systems begins to melt.

Thus, the aim of part I is to present a strong rebuttal to claims that much of our behavior is determined outside of awareness. We will show that many high-profile examples of unconscious influences evaporate once scrutinized, or at least admit alternative explanations that do not require the invocation of unconscious processes. With this reclaimed landscape of the mind established, we turn to the equally puzzling question of how we, as a discipline but also as a society, got to this point.

Part II presents a path toward a true understanding of mind and behavior and begins by asking how we reached the uncritical acceptance of an all-encompassing unconscious mind. We will pursue a trail of fraud, intrigue, and claims about extrasensory perception in an exposé of some of the fake and pseudoscience that has contributed to our current state of affairs. Our journey will take us beyond eye-catching findings to examine the nuts and bolts of how we do research—not only in psychology but also across science and medicine.

Revelations of the instability of many published findings in science and medicine over the past decade or so have raised uncomfortable questions about things we thought we knew.[6] These reevaluations have highlighted the importance of replicating scientific findings, a cornerstone of robust science. Perhaps because such replications are not seen as "exciting" or "sexy," they have been sorely neglected in psychological research. Finding the same effect as somebody else has or redoing the same experiment isn't going to lead to a splashy headline, a high-profile publication, or a TED-talk, but it is, of course, imperative if we hope to build our scientific knowledge on firm foundations. As other commentators have noted, we need to think of each new finding not as a definitive answer but as a single datum in a much larger web of interconnected findings. When we take this more "metalevel" or "meta-analytic" view, we get a much more coherent picture of findings that we should retain and include in our body of psychological knowledge and findings that we can safely disregard as spurious and false.[7]

Our exploration forces us to consider questions that are fundamental for assessing science's role in society: How should science be funded? How should scientists be incentivized and rewarded? How should the outcomes of research be made available to the public? Are current publishing models broken? At first blush, such questions might suggest navel-gazing by researchers. Why should anyone apart from scientists themselves care about scientific publishing practices? In fact, such issues have never been more central to the fabric of our society. We are in an era in which information and data reign supreme; it's never been easier to fact-check, but also never harder to know if the "facts" one discovers are real or "alternative." Sound science is the ultimate fact checker—whether it be psychological science, climate science, or vaccine development—and so knowing how to evaluate, how to consume the science around us, is crucial if we are to maintain science's central role in guiding society.

The overarching theme here might appear negative—debunking, exploding myths, separating the wheat from the chaff—but our goal is a positive one: to reignite and (re)engender confidence in psychological science in particular and science in general. The scientific method is the best one we have for understanding our world, and when it is applied properly and effectively to understanding our own minds, the potential insights and benefits to society are immense. Psychology has already contributed enormously across a wide range of domains. In fact, it is currently enjoying a heyday in the application of behavioral science principles in all spheres of life.[8] We are excited and optimistic about its future, but to ensure we realize the potential, we need to reclaim the science of the mind.

What Is Consciousness?

This book is about the shaky foundations of the idea of a smart unconscious. So what do we mean by the term *unconscious*? Being an absence of something, it is best unpacked by considering the sorts of things that are commonly described as "conscious" or the senses of the word *consciousness*. This is not a straightforward undertaking: the concept has several meanings. One set of meanings is related to levels or altered states of consciousness. States such as sleep, hypnosis, coma, delirium, intoxication, mindfulness, and hypervigilance describe conditions in which we are less (sleep) or more (hypervigilance) aware of our surroundings than normal. As states, these endure over some extended period of time (at least a few minutes) and hence are distinguishable from momentary experiences such as being conscious of a pain when touching a hot saucepan. We can see this distinction in the linguistic use of the word *conscious*, where we talk about just being conscious or about being conscious *of* something. It is used intransitively to refer to states, as in, "The patient was conscious," or transitively to refer to particular experiences, as in, "She was conscious of the loud drilling outside her window."

Another meaning of consciousness relates to our knowledge of ourselves—self-consciousness or self-awareness. We know that we tend to get anxious at parties, that we enjoy sports, and that what goes on in other people's minds is different from what goes on in our own. Curious experiments, in which a mark is placed on an animal's face and its subsequent behavior in front of a mirror is observed, suggest that many species, such as chimpanzees, have some concept of the self and some sense of self-awareness.[9] And finally, we talk about conscious experiences, such as our perceptions (seeing yellow, hearing a violin), bodily sensations (pain, hunger), and emotions and moods (anger, boredom).

Common to the above uses of the term *conscious* is the idea of knowledge and awareness, as when we look at a clock and say, "I'm conscious of the time." The difference between consciousness and awareness is subtle, and we often use them interchangeably. Most dictionaries define *awareness* as being either conscious of or knowing something. But *awareness* doesn't quite have the baggage that *consciousness* does; we tend to talk of altered states of consciousness rather than altered states of awareness, for example.

The association between consciousness and knowledge is also subtle. For many aspects of knowledge, saying, "I know such-and-such," and, "I'm conscious [or aware] of such-and-such," is virtually the same thing. But there are many things we know, such as skills and abilities, where this equivalence breaks down. "I know how to tell a burgundy from a claret" and "I'm aware of how to tell a burgundy from a claret" are very different, as are, "I know how to hit a backhand," and, "I'm aware of how to hit a backhand." Thus, the sorts of things we know—in the sense of being conscious of—are generally taken to be things we can report on. The ability to report or describe some fact about ourselves or the world is the signature of us being conscious or aware of that fact, and reportability is crucial because it's something that an external observer can record. Note, importantly, that we talk about being able to report something rather than simply being able to describe it using language. Language is certainly sufficient to report our awareness of many things ("I know that it's midday"), but it is not necessary. If you're asked to stand up when it's midday, then by standing up, you are reporting your awareness that it's midday but doing so without the use of language. In effect you are reporting, "I understand the instructions you gave me and am standing up to indicate that it's midday." Thus, our ability to demonstrate possession of many types of knowledge can be achieved either linguistically or via a nonlinguistic voluntary report.[10]

The sense of consciousness that's most important for us in this book is the sense in which it refers to states that can be reported, and conversely, unconscious means ones that cannot be.

The Unconsciousness of Mental States

An argument is often made to the effect that there necessarily must be unconscious processes going on in the brain. It points out that there must be such processes because we lack awareness of many events going on in our own brains (for instance, in the visual system). The neural processes by which signals are registered at the retina, transduced along the optic fiber, and decoded in the visual cortex according to properties such as color, shape, and movement are completely unconscious. Indeed, in this sense, we lack awareness of all brain processes, apart from such things as headaches and light sensitivity. This type of argument would obviate the need for any kind of examination of the empirical evidence for unconscious mental processes.[11]

But this argument confuses two different senses of the term *unconscious*. There can be unconscious mental events or states and unconscious nonmental ones. The latter are not particularly contentious. Everything that goes on in a car engine is unconscious, but only in the uninteresting sense that a piece of physical machinery is not the sort of thing that could be conscious. A car engine lacks consciousness in the same way that it lacks bravery. Likewise, the fact that we are unconscious of most of the neurophysiological processes that take place in our brains is uncontroversial. Things become much more interesting when we focus on mental states, such as thoughts, emotions, beliefs, perceptions, and desires—states that represent something in the world—and ask whether such events can cause behavior even if they are unconscious.[12]

The standard view of the relationship between brain events and mental events, known as *functionalism*, views it as akin to the relationship between hardware and software in a computer. These are not fundamentally different things: the hardware is what realizes or causes the events taking place in the software. Imagine that a computer spreadsheet is doing some arithmetic—it's working out 27×56. It is a correct and sufficient description to say that it's manipulating symbols according to a set of rules and the meanings of various functions such as multiplication. This explanation is in no way invalidated by the fact that underneath the spreadsheet, billions of electrical events are going on that don't feature in the higher-level explanation. By the same token, the existence of unconscious neural, nonmental events in the visual system is irrelevant to the issue of whether our behavior is best explained with or without reference to unconscious mental states. This can only be established by the normal methods of scientific observation and experiment.

Moreover, there is another fallacy with the claim that there necessarily must be unconscious processes going on in the brain. The claim rests on an overly bottom-up view of brain processes, in which information flows in only one direction. The senses register energy such as light falling on the retina; this is transmitted to brain areas that decode color, shape, movement, and so on; and then eventually information reaches "higher" brain areas where the meaning of the object in front of us (a picture of Barack Obama, for instance) is finally determined. On this view, we become aware of information computed only in the later stages of this processing pipeline, with everything preceding that being unconscious. But it is a gross mistake to think of the linkage between brain activity and consciousness as one-directional and purely bottom-up. The relationship is much more intimate than this, and indeed there are remarkable examples of top-down influences of consciousness on brain processing. For example, recent neuroscience experiments on individuals with electrodes implanted in their brains (for the assessment of epilepsy) have shown that we can exert conscious, volitional control of single neurons. We can consciously decide to make individual neurons in our own brains fire in both sensory and higher-level brain regions.[13]

Although the concept of the unconscious has a long history going back at least as far as the Swiss physician Paracelsus (1494–1541) and includes a substantial treatment by Gottfried Wilhelm Leibniz in his book *New Essays on Human Understanding* (finished in 1704 but not published until 1765), it is most commonly associated with the work of Sigmund Freud (1856–1939). Freud believed that unconscious motivations and emotions taking place below the surface of the mind play a significant role in human behavior. For Freud, unconscious mental states are not simply those states of which we are unaware; they are a subset comprising socially unacceptable urges and desires, as well as traumatic memories, all of which we actively repress from reaching consciousness. Although they are not introspectively accessible and cannot be explicitly reported, techniques such as free association and dream analysis allow them to be revealed.

The modern fields of psychology and psychotherapy have not been particularly kind to Freud's views, often regarding psychoanalysis as a pseudoscience.[14] But it is worth distinguishing between his conception of the unconscious on the one hand and his empirical methods on the other. Today we tend to regard dream interpretation as akin to storytelling: the psychoanalyst provides an interpretation of the dream in terms of psychoanalytic theory, but since this is necessarily done after the fact, such interpretations are untestable. The hallmark of scientific theories is their ability to make potentially falsifiable predictions, something that psychoanalysis struggles to do (and even when it does, they rarely emerge in credit when tested). None of this means that the concept of the unconscious is itself incoherent, although modern conceptions of the unconscious are rather more general than those accepted by Freud. For example, we have no difficulty today in accepting the possibility of unconscious perception, in which events and objects in the environment around us can in theory influence our behavior even when we are unaware of them (as in subliminal perception).

Reportable, conscious states seem to underlie many of our decisions and behaviors. To state the obvious, we often do things because we have conscious reasons for doing them. This may seem trivial but is nonetheless worth dwelling on, not least because many influential schools of thought—most notably behaviorism—have been deeply skeptical about drawing straightforward links between conscious reports and behavior. Some of this skepticism is partly justified. For example, we need to be very careful about the reactive effects of verbal reports: the very act of giving a description of our mental states may change those states.[15] However, modern theories and methods make an emphatic case for the importance of conscious knowledge, beliefs, and attitudes in determining behavior.

This is exemplified in the theory of planned behavior (TPB), one of the most influential and truly deep theories in all of the behavioral sciences.[16] TPB offers a deceptively simple explanation for why we act as we do. It says that only two things matter: our intentions and our control. When we have a strong intention to engage in a behavior and perceive that we have control over that behavior, then we will do it. Take smoking cessation as an example: people who intend to give up smoking and perceive that they have control over whether they smoke will give up, whereas those without either the intention or the perception of control (or both) will not.[17] The theory in addition says that intentions come from three sources. First, we are very strongly influenced by subjective norms, our internalization of the views of other people whose opinions matter to us. If your partner's views matter to you and your partner takes a dim view of your smoking, you're more likely to form the intention to give up. Second, our attitudes to the behavior are important. The better you feel about giving up smoking, the more likely you are to form the intention to give up. Finally, your estimation of how easy or difficult it will be to give up will also influence your intention. This sense of control is the same factor mentioned above: it is assumed to have a direct impact on behavior as well as an indirect impact via intentions.

According to TPB, it is only these factors, and no others, that matter. You might object, for example, that surely personality matters. Aren't extroverts less likely to quit smoking than introverts? The theory doesn't deny that such factors could be associated with giving up smoking, but it says that the only pathways by which they can do so are via subjective norms, attitudes, or perceived control. Thus, if extroverts do indeed struggle to stop smoking, it must be because they feel less pressure from other people's views, or they have a less favorable attitude to stopping smoking, or they do not feel they have adequate control.

The precise details of the theory are not crucial; what matters is that the theory is eminently testable (and indeed it has survived more or less intact despite thirty years of extensive testing) and that all its ingredients are conscious, reportable aspects of behavior.[18] When researchers set out to test the theory, they do so by constructing questionnaires comprising many items, all of which attempt to measure different aspects of the person's subjective norms, attitudes, and perceived control. Each item (such as *Most people who are important to me think I should give up smoking in the next 12 months*) is accompanied by a rating scale on which the respondent indicates their agreement or disagreement with the statement, and the different items under each type are averaged to give a single measure of the three predictors for that person. Then an analysis is conducted to determine whether those individuals with strong subjective norms, attitudes, and perceived control for quitting smoking are indeed more likely to quit. Clearly, the questionnaires elicit conscious, reportable beliefs and attitudes; this is a theory that fundamentally attributes behavior to a combination of fully conscious factors. The theory *could* be supplemented by additional unconscious factors, though these would of course have to be measured by some means other than explicit reports (the implicit association test, discussed at length in chapter 5, would be an example). What is remarkable is that TPB has achieved the explanatory successes that it has even without the addition of unconscious factors (the jury is still out on whether there are domains in which such factors would extend its power even more).[19]

A final distinction, between access (A) consciousness and phenomenal (P) consciousness, deserves some mention. Two things are going on when a state is reportable. One is that the state exists in such a form or has reached the necessary level of internal strength that it can be turned into a report via the relevant mental apparatus. The state is accessible, in other words, to that apparatus. The other is that the state has a subjective "feel" to it, that there is something it's like to experience that state. Think of being asked to report the level of pain (between 0 and 10) you feel to different stimulations. When pricked on the finger, you say "2." The pain you consciously feel connects to the reporting apparatus in your brain and causes you to make this report; the pain is accessible to that apparatus. At the same time, the pain feels a certain way to you and has certain *qualia*: it is very sharp and unpleasant, not at all like heat, but brief. This is the pain's subjective feel, its phenomenology (a term philosophers favor to refer to the way we experience things).

This distinction between access and phenomenal consciousness is a profoundly deep one. Broadly speaking, we have a very good understanding of the former and not the least inkling about the latter. The sort of machinery that is necessary to enable states to be accessible is well understood. When we run a spreadsheet on our computer and ask it to perform a complex arithmetic calculation, the result that it prints on the screen is a "report" on the spreadsheet's calculation. That calculation is accessible to the function that prints the output. But we don't imagine for one moment that the computer has any conscious experience associated with generating the answer; we do not attribute to the computer any subjective what-it-is-like-ness to do so. The philosopher Thomas Nagel famously reflected on what it's like to be a bat. Perhaps the bat has some conscious experience of the world around it, as revealed by its echolocation system, an experience no doubt profoundly different from our own familiar subjective experiences of color, taste, touch, and so on.[20]

The mystery of P-consciousness is one of the most fundamental unresolved problems in all of science, and not for nothing has it been given the label the "hard problem." We simply have no idea what the processes are in the brain that generate the experiences of redness, saltiness, euphoria, and so on. We can build a computer that tells colors apart just as accurately as the human eye, but we have no conception of why the human (and perhaps chimpanzee, dog) brain, but (presumably) not the computer, experiences colors subjectively in the way it does. As the famous nineteenth-century biologist T. H. Huxley (1866) asked, "How it is that anything so remarkable as a state of consciousness comes about as the result of irritating nervous tissue, is just as unaccountable as the appearance of Djin when Aladdin rubbed his lamp."[21] But the good news is that for the purposes of this book, we don't need to solve this hard problem. Our focus is on determining whether there are any major mental processes that cause or influence our behavior without being A-conscious—that is, without being reportable.

Consciousness and the Brain

However mysterious it is that states of consciousness come about as a result of irritating nervous tissue, the brain obviously is the seat of consciousness. Neuroscience has made enormous strides in understanding the linkage between brain activity and conscious experiences and has wholeheartedly rejected dualism—the claim that the mind cannot in any sense be understood in terms of the physical world, of which the brain is a part. There are any number of examples of this linkage. In pioneering research in the 1960s, the physiologists G. S. Brindley and W. S. Lewin were able to induce the conscious experience of a small, bright dot in a very specific part of the visual field by electrically stimulating a corresponding part of the visual cortex in the occipital lobe of the brain. These bright dots (called "phosphenes") were always experienced at the same location for a given point of brain stimulation, and nearby stimulation sites induced nearby phosphenes.[22] It is hard to imagine a more compelling linkage between brain activity and conscious experience.

In the past two or three decades, neuroscience has provided tools for establishing brain-consciousness linkages vastly more complex than Brindley and Lewin's demonstration. Functional magnetic resonance imaging (fMRI) can be used to measure the level of activity in distinct parts of the brain. The method achieves this by detecting the amount of oxygen in the blood, which increases when a brain region becomes active. The technique has been used to determine, entirely noninvasively, whether an experimental participant is thinking about a chair or a shoe. Although the pattern of neural activity is never quite the same on two occasions when you're thinking about a chair, this activity is sufficiently different from the activity evoked by thinking about a shoe that it can be measured by fMRI and correctly classified as representing conscious thought about one or the other.[23] This method has been used to detect consciousness in individuals in a vegetative state following head trauma. Imaging via fMRI was able to detect different patterns of brain activity in one woman when she was asked to imagine playing a game of tennis versus when asked to imagine walking around the rooms of her house. Although she was unable to give any behavioral indication that she understood what she was being asked to do, her brain reacted in a way that strongly suggested consciousness.[24]

Several theories have been proposed in recent years attempting to explain the relationship between consciousness and the brain in more detail. The most influential of these is global workspace theory (GWT). One of the key ideas in this theory is that (access) consciousness arises when the intensity of activity in a particular brain module exceeds a threshold, at which point it becomes amplified and broadcast to the central controller—the global workspace—and to other modules. A distant object may not be consciously registered until it gets closer, at which point "ignition" takes place: the object is now consciously perceived, and its existence becomes shared with other brain systems such as the ones that control movement (hence enabling avoidance of a collision) and speech (enabling issuing a warning to others). Ignition involves the allocation of attention to the object and the sharing of information globally across the brain. The most recent versions of GWT suggest that the global workspace comprises interconnected neurons in frontal, parietal, and anterior temporal brain areas, connected to the specialized peripheral modules via bidirectional links. The theory predicts, therefore, that the sorts of imaging techniques described above should reveal conscious experiences in frontoparietal areas, with long-distance connections between these areas being fundamental to consciousness.[25]

Despite this progress in understanding the brain basis of consciousness, it is fair to say that we are still very far from being able to explain consciousness in terms of brain machinery. Indeed, as far as phenomenal consciousness is concerned, its relationship to the brain is as mysterious now as it was in Descartes' time. For access consciousness, experimental tests of GWT have proved disappointing. In contrast to the theory's claim that consciousness depends on frontal and parietal systems, virtually every part of the brain seems to have some linkage to consciousness.[26] Given Brindley and Lewin's ability to induce conscious phosphene experiences by stimulation of the visual cortex, perhaps this should not surprise us. It may be the case that for almost every brain region, whatever the particular set of functions it contributes to our mental life, it is also necessary for conscious awareness of that function. So the visual cortex is necessary for visual consciousness, auditory cortex for auditory consciousness, and so on. In this sense there is no "Cartesian Theater" (the philosopher Daniel Dennett's evocative phrase), a single location where consciousness happens like a movie being projected onto a screen.

* * *

The nature of consciousness is one of the most perplexing problems in all of modern science and philosophy. But we don't need to wait for a solution to this problem before we can ask what the role and scope of *unconscious* mental processes is. This is the question we address in the following chapters.

2 Moving, Deciding, and Free Will

Our starting point for reflecting on the role of the unconscious in our decisions and behavior is to examine very simple actions and skills, such as catching a ball or pressing a button. By focusing on uncomplicated behaviors and movements such as these, we should be able to get a clear sense of what it would mean for them to be made or controlled unconsciously. But as we will see, it is not obvious that unconscious mental processes in fact play a large role in the execution of simple movements and skills.

Simple Actions and Brain Precursors of Decision Making

Except in some psychiatric conditions such as schizophrenia, everyone has the sense that they are free agents consciously deciding how to act. Yet in one of the most thought-provoking controversies in modern neuroscience, the notion of free will and, in particular, the intuitive assumption that it is our conscious thoughts that cause our actions came under considerable duress following famous experiments conducted by the neurophysiologist (and inaugural winner of the sadly now-defunct Virtual Nobel Prize in Psychology) Benjamin Libet and his colleagues.[1] They monitored electroencephalographic activity on the surface of the scalp while participants pressed a key or flexed their finger. Electroencephalography (EEG) is the technique of recording electrical brain activity and attempting to infer the underlying brain processes causing that activity. In Libet's procedure, participants freely chose when to make a voluntary movement. The crucial twist in the experiment is that they were also asked to report the time point at which they felt the intention to move. Participants observed a spot rotating on a clock (illustrated in figure 2.1) and made their timing reports by

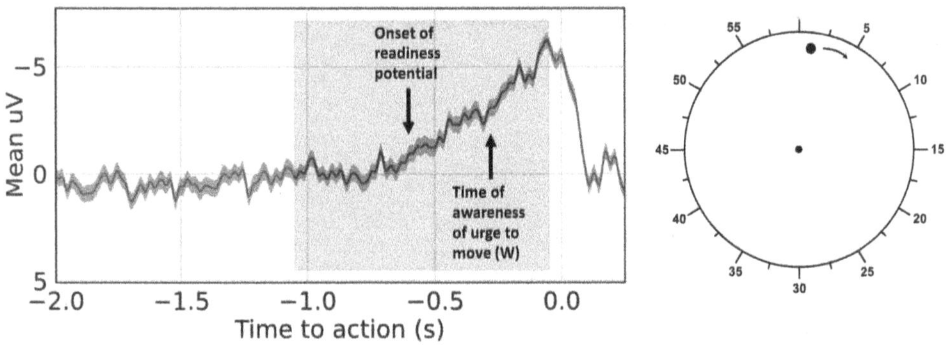

Figure 2.1
Schematic illustration of Benjamin Libet's famous experiment on the timing of the conscious experience of willing a voluntary movement. The graph shows the "readiness potential" on the vertical axis (in negative microvolts) across a short period of time (horizontal axis). The readiness potential is electrical activity measurable at the surface of the brain that is caused by the motor cortex as it prepares a voluntary action. In Libet's procedure, participants freely choose when to make a simple action such as pressing a button, and the point at which this effect occurs is marked as time zero. The graph shows that neural activity begins approximately half a second before the button is pressed. Simultaneously, participants are watching a spot moving around a clock face (illustrated on the right; the spot rotates much faster than a normal second hand) and report the time at which they feel the conscious will to press the button (W). The point at which this feeling is temporally located is appreciably later than the onset of the readiness potential.

observing the dot's location at the point of becoming conscious of their urge to move (these are called "will" or "W" judgments).

Strikingly, Libet found that, to put it colloquially, the brain "decided" to move before the individual did. These timing judgments followed rather than preceded the first neural marker of movement intention, the readiness potential, and indeed the time interval between these could be as much as a second (see figure 2.1; a recent meta-analysis estimated this interval as 0.48 seconds on average).[2] Libet and many subsequent commentators have taken these results as evidence that conscious intentions do not cause voluntary actions but are instead epiphenomenal or secondary effects of the true, unconscious causes of such actions, namely, neural events.

Other research has extended the method using recordings of activity in single neurons in the medial frontal cortex part of the brain, which show progressive recruitment over several hundred milliseconds prior to participants' reported experience of the urge to move.[3] The same basic finding has also been obtained in a modern neuroimaging adaptation of the Libet task in which participants watched a stream of letters (one letter every half second) and made a left or right button press at a freely chosen time.[4] They then reported the letter that had been on the display at the moment they felt they formed their conscious choice. On the basis of advanced methods for decoding neural activity, it was found that several seconds before the choice was made, and long before it was conscious, two brain regions (frontopolar and precuneus/posterior cingulate) contained information that predicted that choice, suggesting that unconscious processes make a significant contribution to decision making.

Libet's research and subsequent extensions have been the subject of a vast psychological and philosophical literature that explores the concept of free will and its relation to the brain. The key issue is whether there are truly brain processes unfolding before any conscious intention to act and, if there are, how those processes should be understood. There are a number of reasons to be very wary about the interpretation that Libet and others have offered for their findings.[5]

One particularly noteworthy discovery, confirmed in subsequent studies, is that when lateralized readiness potentials are measured instead of standard potentials, the key Libet effect is not obtained. The lateralized readiness potential reflects brain activity that is specific to the arm that is making the movement, bearing in mind that each half of the body is controlled by the motor cortex on the opposite side of the brain. It is a more appropriate indicator of hand-specific movement preparation than the readiness potential, which is at least in part a marker of very general preparation for a future movement. W judgments tend to precede, not follow, lateralized readiness potentials, suggesting that there is no misalignment between the conscious urge to move and the onset of neural activity associated with that specific movement. Consistent with this, research has also found no evidence that the lateralized readiness potential develops prior to a signal (a tone) in a variant of the Libet task in which participants make spontaneous decisions about which hand to move as soon as they hear the signal, which it would have done if the brain makes unconscious decisions prior to the signal-induced conscious ones. These findings run counter to the claim that readiness potentials are markers of unconscious movement preparation.

Two additional findings confirm that the readiness potential is not—as Libet supposed—an appropriate measure of preparation for action execution.

First, this brain potential is indistinguishable on trials on which participants choose to move versus those on which they choose not to (when given the option). If this potential is a marker of movement preparation, then a necessary prediction is that it should be greater on movement than on nonmovement trials, contrary to what is observed.

Second, the readiness potential signature of movement preparation is virtually eliminated in conditions where participants make voluntary movements in the absence of a clock and with no requirement to report W judgments. The implication of this is that the preparatory neural activity that Libet took as evidence of unconscious movement preparation has more to do with dividing attention and preparing to make a clock judgment. The clock procedure, which was designed to measure mental events, seems in fact to alter the neural activity to which these mental events are related. The readiness potential signature is also eliminated when participants are asked to make deliberate decisions (for example, choosing which of two nonprofit organizations to make a donation to) rather than the arbitrary ones typically studied in Libet-style experiments.[6]

Other research has tended to further highlight complexities in the measurement and interpretation of subjective correlates of willed actions. It has been shown, for example, that W judgments are highly sensitive to postaction factors such as the timing of feedback. When auditory feedback accompanying the participant's movement is delayed, so is the W judgment.[7]

This urge toward caution is further supported by the phenomenology of the Libet task. In performing rapid finger movements, we do not usually have a distinct awareness of wanting to move and then a distinct awareness of moving. Rather, we just have a unitary awareness of the act. By forcing participants to try to time their conscious intentions, the Libet task may unintentionally bias them to report their movement rather than their urge to move. Moreover, reported decision times are highly variable, perhaps accurate to no more than ±300 milliseconds, and under some circumstances as many as 30 to 40 percent of them are implausibly early or late,—for instance, being located after the movement itself.[8]

These observations all relate to empirical questions surrounding the interpretation of the Libet task. A further issue has a more theoretical nature and takes as its starting point the fact that an intrinsic aspect of decision making is that choices are preceded by the accumulation of preference. Decisions are not reached instantaneously out of thin air. Over time, we acquire reasons for acting one way or the other, and this accumulation is no less a feature of simple choices like pressing the left or right button in a laboratory experiment as in complex real-world contexts such as deciding which house to buy.

This seemingly intuitive point is important because it suggests that even if the standard interpretation of the Libet task were correct—that it demonstrates preparatory neural activity prior to the point at which a conscious decision is reported as having occurred—this would not be evidence of unconscious influences on decision making. Libet's assumption that "if a conscious intention or decision to act actually initiates a voluntary event, then the subjective experience of this intention should precede or at least coincide with the onset of the specific cerebral processes that mediate the act" implies that conscious intentions are not brain processes but can nevertheless cause such processes.[9] If, in contrast, we assume that conscious intentions *are* brain processes, then they could only be simultaneous with those processes if the intentions arise instantaneously from the neural processes that instantiate them. Much more plausible is that the time course of the awareness of an intention is gradual and lags behind the earliest mediating brain processes.[10] Think of a chess-playing computer taking several seconds to decide its next move. The time at which the decision is reached would invariably be later than the time at which the electrical activity mediating that decision began, even if the decision is wholly based on explicitly represented reasoning.

It is surely the case that the process of forming a decision takes time. Suppose that a threshold degree of bias or preference (100:0) is required before a person makes a voluntary movement of the left or right hand. Then the accumulation of bias prior to reaching this threshold could be entirely conscious and neurally measurable for tens or hundreds of milliseconds, even before it compels the button press. When the individual reports the time at which they consciously made their decision, perhaps they (perfectly reasonably) report the point at which their bias reached, say, 70:30 rather than the point it first drifted away from 50:50. The key point is that the threshold for detecting neural activity does not have to be the same as the threshold for reporting a state of awareness. All in all, the Libet task and its many variants provide little compelling evidence to believe that unconscious brain processes start to unfold prior to conscious intentions. If anything, they serve to demonstrate the close alignment of awareness and action.

Conscious Will as an Illusion

The folk psychological view that conscious thoughts cause our decisions and behavior faces another major challenge from the substantial body of evidence suggesting that our conscious thoughts are often inferred after the fact. Rather than making conscious choices and immediately and passively experiencing those thoughts, an alternative possibility is that the thoughts are constructions created post hoc and that the true causal work is done by unconscious states of mind and brain. This is the essence of the will-as-illusion viewpoint, which emphasizes that experiencing an intention prior to an action is no guarantee that the intention caused the action. The idea of conscious agency is a "willusion."[11]

In one particular version of this approach, and in contrast to the intuitive view that our decisions and behaviors are caused by conscious intentions, it has been argued, particularly by the social psychologist Daniel Wegner, that they are instead caused by unconscious processes that may simultaneously produce illusory experiences of conscious will.[12] Specifically, it is proposed that unconscious states of mind/brain cause two things: the voluntary action itself and a conscious thought about the action (intention). As a result of the constant conjunction of thought and action, an experience of will is created by illusory inference even though the thought itself is not the true cause of the action. Wegner draws an analogy with a ship's compass. Someone looking at the compass and relating it to the ship's course might form the impression that the compass is actually steering the ship, yet we know that the compass exerts no such control over the ship's movement. The compass reading is an effect, not a cause, of the ship's course, which is in fact caused by a whole raft of separate factors and processes, such as the prevailing wind and the position of the ship's wheel and rudder.

Wegner relates mental causation to the classical theory of physical causation, famously proposed by the philosopher David Hume, which proposes that the constant conjunction of two events in the world (object A hitting object B and making it move) creates the conditions for us to infer that object A caused object B to move.[13] For Hume, there is nothing about causation that can be directly perceived. Instead, we infer that two things are causally linked on the basis of their being repeatedly paired and in "constant conjunction." Constant conjunction in Wegner's theory can be broken down into three features: *priority* (the thought should occur just before the action), *consistency* (the thought should be compatible with the action), and *exclusivity* (the thought should not be accompanied by other potential causes).

Wegner's principal support for this theory comes from demonstrations that illusions of will can be created in which people either experience will when their conscious thoughts are objectively not the cause of their actions or fail to experience will when they objectively are. For example, a collaborative computer environment was created in one experiment that is so famous that it has its own name—the I Spy study.[14] The experimental participant and a confederate jointly controlled a mouse that moved a cursor around a complex display comprising numerous objects. On the critical trials, the participant heard a word over headphones that referred to an object on the display (for example, a swan). If the confederate arranged for the cursor to stop on the object immediately after the word was presented, then the participant tended to report that she intended the cursor to stop. Thus, the occurrence of the object's name and a movement directed to that object induced a sense of will even though the participant was not responsible for the action.

In another influential study, the Helping Hands experiment, participants watched themselves in a mirror with their arms out of view by their sides while a confederate stood behind them (figure 2.2).[15] The confederate's arms were extended forward to where the participant's arms would normally be, and these arms performed various actions such as giving an "okay" sign. When the participants heard instructions over headphones previewing each of these actions, they judged that they had greater control over the arms movements. Wegner has concluded from such demonstrations that the experience of conscious will is an illusion in the same sense that the experience of physical causation is. In both cases, our minds draw inferences when the conditions are appropriate, namely, when constant conjunction is present.

There have been numerous responses to Wegner's radical position on will and the conscious causation of behavior.[16] One noteworthy observation is that these experiments do not induce anything remotely resembling full-scale experiences of agency.[17] In the Helping Hands study, for example, participants rated their sense of vicarious control on seven-point scales (with 1 = not at all and 7 = very much). Although participants reported a significantly stronger feeling of control when the actions were previewed

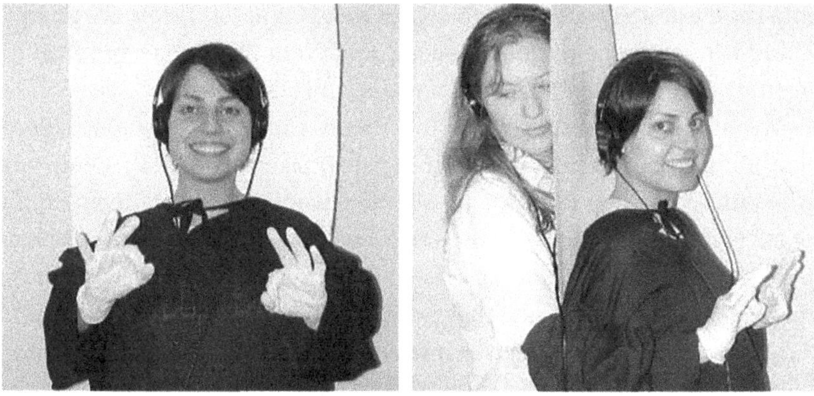

Figure 2.2
Daniel Wegner and colleagues' Helping Hands experiment. The participant appears when viewed from the front to be using her arms normally (left). However, in fact (right) a confederate's arms are extended forward and the participant's arms are out of view by her side. The participant watches the effects of various arm and hand movements in a mirror and makes agency judgments about her degree of control over these actions.

auditorily, their average ratings were never greater than 3 on this scale. Hence, it can hardly be claimed that they reported experiencing a feeling of control over the confederate's actions.

The appropriateness of the analogy to physical causation and the relevance of Hume's principles have also been questioned.[18] We often experience will even when an intention precedes an action by a long interval, such as when making the decision to go on a vacation at some time in the future. The analogy with physical causation is curious because the conclusions drawn in the two cases seem very different. In the case of physical causation, even if it is accepted that our knowledge of causation is an inference based on constant conjunction and that we can in consequence experience illusions of causation, most people do not conclude that physical causation itself is a fiction or that perception is generally illusory. Rather, we conclude that there are real causal connections in the world but that our knowledge of them is indirect and largely inferential.[19] In contrast, on the basis of illusions of agency and will, Wegner's conclusion is that free will and the conscious causation of behavior are illusions. The illusions per se cannot prove this. They merely show that we lack direct access to linkages between thought and action.

Finally, it must be borne in mind that retrospective memory distortions can create false reports concerning experiences of will and we must therefore treat such retrospective reports with caution. This point is well illustrated by the "choice blindness" experiments of Petter Johansson and his colleagues.[20] They asked their participants to choose which of two photographed faces was more attractive. By employing a subtle card trick, these researchers were able on some trials to present the rejected face as a prompt for the participant to explain his or her choice. Detection of this manipulation was rare, and, strikingly, participants readily gave justifications for their choice of this face, even though it was not the one they had chosen. Moreover, participants tended to misremember having chosen the rejected face on the manipulated trials. Thus, in the space of a few seconds, memory can rapidly distort recollection of a choice and, presumably, the reasons for it.

Perceptual-Motor Skills

Whatever the status of unconscious influences on simple voluntary actions, most would agree that it is in the domain of perceptual-motor skills like catching a ball or riding a bicycle that such influences will be most easily demonstrated. After all, our subjective experience is that we do not need to "think" in order to carry out these skills once we have learned them. Yet in achieving success in such skills, we implicitly demonstrate knowledge of complex physical laws despite lack of awareness of those laws. We must therefore probe more deeply into such skills and people's conscious knowledge of them.

Although there have been relevant studies on numerous skills such as typing, playing a musical instrument, and sports skills including golf putting and table tennis, we will focus on ball catching, a difficult ability that has been the subject of several research studies over the past few years. Because this work has been targeted both on understanding the decision-making cues that people use in order to catch balls as well as their awareness of those cues and of the basis of their expertise, it provides a major test case of the role of unconscious mental processes in behavior.

In the general case of catching an object, the catcher has to make decisions about forward or backward and lateral movement and also has to take account of nonstandard trajectories. The catcher might, for example, have to run not only forward to intercept a ball but also to the right, and at the same time the interception point will differ for a ball and a Frisbee, as the latter generates lift that affects its flight. For this general case, the algorithm that people employ is not well understood. However, for the restricted case in which no lateral movement is required because the catcher is standing in the object's plane of motion (the object is thrown directly at the catcher, the only variable being whether it will fall short, hit the catcher, or go overhead) and the object is on a ballistic trajectory (resulting from gravity and air resistance), the algorithm is well characterized.

The explanation of catching skill under these conditions, known as the *Chapman strategy*, assumes that the catcher's behavior depends on a single variable, α, the angle of eye gaze with the horizontal when the catcher looks at the ball in flight, as illustrated in figure 2.3.[21] Initially, α increases in the early part of the trajectory as the ball rises. If α continues to increase, the ball will fall behind the catcher, who therefore must run backward to intercept it. If α begins to decrease, then the ball will fall in front of the catcher, who therefore must run forward to intercept it. Only if α increases at a decelerating rate will the ball fall directly at the catcher's location, hitting her in the eye. Thus, monitoring α is sufficient to guide successful ball catching under the constraints of no lateral movement and a ballistic trajectory. By moving forward or backward in such a way as to generate a value of α that increases at a decreasing rate, the person will converge on the ball's landing point. Generally experimental tests confirm that the theory provides a good description of people's actual behavior.[22] Our question will be whether people are unaware of the way in which α is the controlling factor in their skillful behavior. People might be entirely unaware that gaze angle is the cue influencing their behavior, or they might be aware of the cue but have an incorrect theory about how they use gaze angle. Or they might have some insight into their use of gaze angle.

The most comprehensive effort to answer this question was undertaken by Nick Reed, Peter McLeod, and Zoltan Dienes[23] and we will consider their findings at some length. They took a simple initial approach to determine whether people's knowledge of their eye gaze during ball catching is accurate and in accordance with the Chapman strategy. They presented participants with scenarios corresponding to cases in which the ball would fall short and they would have to move forward, in which case they could catch it without moving, and in another case in which it would pass overhead and they would have to move backward. For each case, the participants were

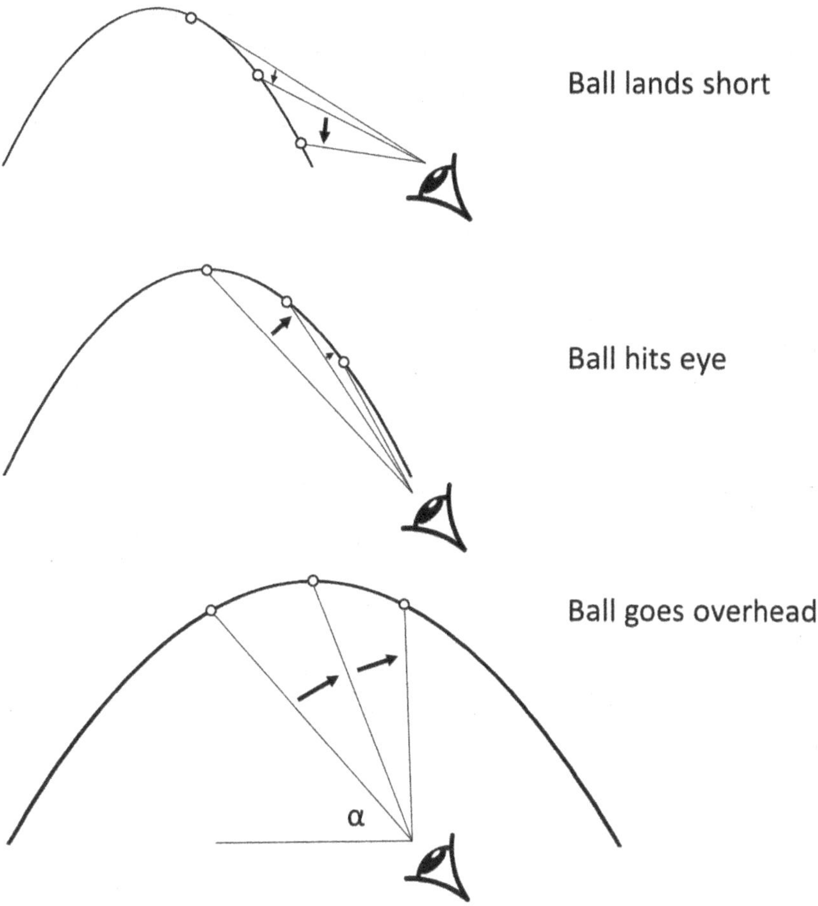

Figure 2.3

Illustration of the way in which the angle of eye gaze, α, changes during a ball's flight. In all cases, α increases in the first part of the ball's flight as the catcher's gaze follows the ball's upward trajectory. In the top panel, the ball will land short of the catcher, and α reduces as the ball falls to the ground. In the middle panel, where the ball will hit the observer in the eye, α increases steadily at a decelerating rate. Counterintuitively, it never decreases. In the bottom panel, the ball will fly over the catcher's head, and in this case α continues to increase. Hence, the catcher can intercept the ball by moving forward or backward in such a way that α increases at a decelerating rate. This is the Chapman strategy.

asked how they would know that they were in the right place or would have to move. Almost none of the participants mentioned the change in angle of gaze. Others referred, incorrectly, to the apex of the ball's trajectory as the key signal. Thus, at first glance, this appears to indicate a profound lack of awareness of ball catching.

The unconstrained nature of the question posed to participants ("How would you know that you were in the right place or that you would have to move?") should raise concerns, however (later in this chapter we will be more explicit about the criteria that should be taken into account when deciding whether a test of awareness is a good one). For example, participants might be unwilling to report low-confidence knowledge or might prefer (despite being aware of the gaze signal) to report a simpler naive theory. Mindful of these issues, Reed and colleagues conducted two further studies with more constrained question formats designed to circumvent them. In their second study, participants were specifically asked how their gaze would change for balls landing short, being caught at knee level, eye level, or overhead or flying overhead. Angle of gaze was explained graphically, and examples of how gaze would change when watching a rocket or a parachutist were described. The results of this study were rather different from those of the first study, although still providing some evidence of unconscious control. Specifically, participants' descriptions were qualitatively correct for four of the five scenarios. Only in the case of a ball reaching them at eye level did participants frequently give incorrect reports about the way gaze would change. As the researchers themselves concede, participants can accurately report how their gaze would change for all flight trajectories except for a ball caught at eye level.

This outcome is particularly interesting because the true dynamics of gaze for a ball that is heading for the eye are not at all intuitive. A common misconception, reported by many of the participants, is that angle of gaze first increases and then decreases. This is incorrect because under such circumstances, the ball is never at a higher elevation in terms of α than it is just as it hits the eye (see figure 2.3). Indeed, video recordings of eye gaze confirm that the angle never decreases for balls caught at eye level. People presumably are prone to the cognitive error called attribute substitution, that is, replacing something that comes to mind easily (the ball itself goes up and down) for something that does not (gaze angle) and concluding erroneously that the latter also goes up and down.[24]

In neither of the studies did participants actually catch balls while reporting their conscious beliefs. It is possible that people are aware of the critical signal (the change in α) while catching balls but unable to recall it out of context. In order to prime access to conscious knowledge as fully as possible, in their final study Reed and his colleagues gave their participants several opportunities to catch balls while thinking about their eye gaze. For each catch, participants made a forced-choice decision among various descriptions of gaze angle (they also made this choice prior to catching balls). The options included statements such as *continuously down, up and then down, up and then remaining constant, up at a decreasing rate* (the correct choice for balls landing at eye level), and so on. The results of this study are crucial to the evaluation of the role of unconscious processes in behavior because considerable effort was made to test conscious knowledge in a context where the skill itself was being displayed and using a sensitive testing method. What were the results?

First, participants appeared to be aware of gaze change for balls falling below or above eye level, as in the second study. For the four cases where the ball is falling below or above eye level, choice of the correct description was generally high. Second, the forced-choice test led to an improvement in accuracy even in the case where knowledge was probed prior to catching balls. This suggests participants can access their eye gaze dynamics. Third, accuracy in the condition where the ball reaches the participant at eye level improved further still when a forced choice was made after an actual catch. This key finding suggests that on many trials, participants were able to correctly access their eye gaze trajectory.

These results lead to a rather different interpretation of awareness in ball catching. We have already seen that the evidence that this skill is unconscious is restricted to situations with no lateral movement and ballistic object trajectories. It is further restricted to cases where the ball reaches the individual at eye level: for cases where the ball will reach the (unmoving) person below or above eye level, insight concerning gaze is accurate. The final set of results suggests that even in this restricted case, lack of awareness is the exception rather than the rule. Put differently, for the vast majority of occasions on which people catch balls—excluding only the rare cases where the ball is precisely converging on the eye—people seem to be aware of the signal that guides their behavior (whether to move forward or backward).[25]

Far from demonstrating that people's reports about their ability to catch a ball are typically erroneous or that they are unable to gain conscious access to the mental computations guiding their ball catching, the results of this important study reveal that the use of gaze angle can be consciously accessed and reported. This is not to say that in the normal course of performing a simple perceptual-motor skill, people are fully conscious of all the relevant mediating mental operations. This is plainly not the case; such intuitive skills are achieved with very shallow phenomenological experience, and we allocate very little attention to and engage in minimal monitoring of those mental operations. Instead, we draw a more modest conclusion: the data do not establish the existence of influences that are outside awareness in such skills. Evidence concerning the simple decision about whether to advance or retreat in order to intercept and catch a ball falls far short of demonstrating independence from conscious control.

A perfectly reasonable response to the ball-catching evidence is that it concerns an established skill: how we behave and what we know about our behavior when performing a skill that we have practiced thousands of times since infancy. What about skill acquisition? A great deal of research has asked whether unconscious processes play an important role in perceptual-motor skill learning—so-called implicit learning—such as learning to putt in golf or adapt to changes brought about by visual distortions. This research is extensive and complex but includes many examples of close linkages during the acquisition of skills between performance and awareness.[26]

For instance, when we have to learn a new mapping between a movement and its outcome, awareness is strongly associated with movement adaptation. Consider the way in which the cursor on your computer screen moves as you move the mouse. Normally the relationship between these is very simple: a movement of the mouse in the forward-backward direction translates into an up-down cursor movement, and a movement of the mouse in the left-right direction translates into a left-right cursor movement. But suppose that a gradual distortion of this mapping was surreptitiously introduced, so that each day you have to move the mouse 1 degree farther clockwise to achieve the same cursor movement. After forty-five days, you have to move the mouse along an axis 45 degrees clockwise from the forward-backward axis in order to move the cursor straight up and down on the screen. People learn to adapt quite well to such distortions, but do they tend to be aware of doing so? Research shows that indeed awareness and adaptation are strongly linked in such situations.[27]

Over the past forty years, researchers have devised numerous laboratory tasks involving simple movements and actions with the aim of examining the alignment or lack of alignment of awareness and behavior. One of the pioneers of this work was the influential Oxford psychologist Donald Broadbent. Usually awareness is assessed in these studies by simple verbal reports. Standing back from specific examples of the findings from such experiments, there is a common thread to the general findings that emerge: evidence of unconscious influences on behavior is followed by more careful tests revealing the opposite, in a recurring cycle.[28]

A researcher invents a new laboratory task and reports an initial finding, often based on a small sample, that participants lack conscious insight into their learning and performance. The lack of conscious insight is a null result—that is, a failure to observe something, in this case meaningful levels of reportable information about the task. As we will discover throughout the course of this book, behavioral science suffers from a pernicious tendency, based on underpowered experiments, to misinterpret the absence of evidence (in this case, failing to detect reportable awareness) as evidence of absence (awareness is conclusively absent), despite the fact that these are different things. Later, other researchers try to replicate the original study with greater power (that is, larger sample sizes) and better tests of awareness that pay greater heed to the criteria noted previously. These more careful experiments typically find contrasting results, namely that awareness about the task is deeper than initially thought. After a while, interest in that particular laboratory task wanes, but before long, a clever and interesting new task is devised and the entire cycle plays out again. What is rarely and perhaps never seen is an experimental demonstration of a misalignment between behavior and awareness that is readily replicable and robust in independent research.

The Curious Case of D.F.

The final two domains that we consider involve research inspired by neuroscience and neuropsychology implying that distinct conscious and unconscious routes exist in the brain for the control of action.

Neuropsychological and behavioral evidence has suggested functional differences between two neural pathways involved in the processing of visual information (see figure 2.4). The ventral perception pathway includes

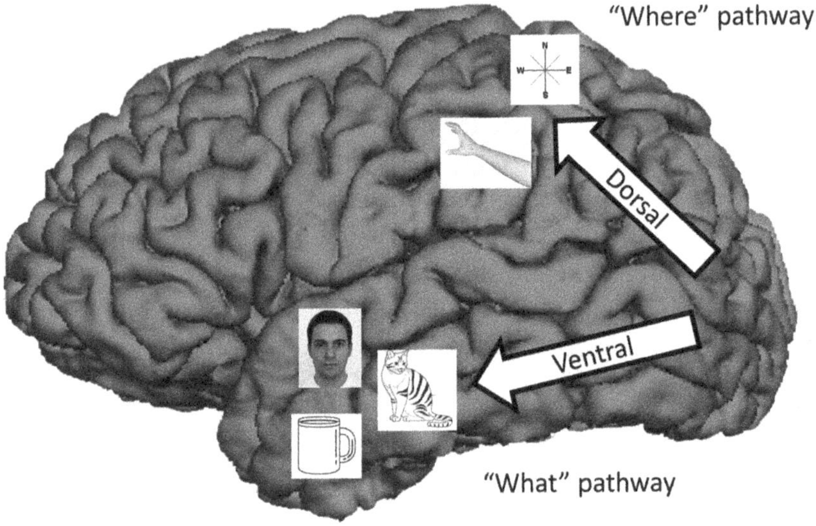

Figure 2.4
The two-streams hypothesis of vision. David Milner and Melvyn Goodale proposed on the basis of anatomical and behavioral evidence that after reaching the occipital lobe at the back of the brain, visual information follows two streams as it undergoes further analysis. The ventral stream (the "what" pathway; "ventral" refers to the front or lower) connects to the temporal lobe and is concerned with object recognition, such as recognizing a cup as a cup. The dorsal stream (the "where" pathway; "dorsal" refers to the back or top) connects to the parietal lobe and processes information about spatial location. Importantly, the hypothesis also proposes that object recognition in the ventral stream is conscious, whereas spatial processing in the dorsal stream is unconscious.

projections from primary visual cortex to inferotemporal cortex, while the dorsal action pathway projects from primary visual cortex to the parietal lobe. Neurons increase the size and complexity of their receptive fields as one moves along these pathways. The ventral stream is often referred to as the "what" stream and the dorsal pathway as the "where" stream on the basis that the former seems to involve processing of object identity while the latter is concerned with spatial awareness and reaching. Of more relevance here is the proposal, due to Melvyn Goodale and David Milner, that object recognition in the ventral stream is conscious whereas computations for the guidance of actions in the dorsal stream are distinct and unconscious.[29]

Goodale, Milner, and their colleagues studied a patient whose behavior seems to provide support for such a dissociation between dorsal and ventral stream processing. This famous (among psychologists) individual, D.F., suffered from agnosia, a condition in which the ability to recognize objects and people is impaired despite apparently normal basic visual and memory abilities. It is commonly the result of brain damage in ventral stream structures, particularly lateral occipital cortex. D.F. had profound difficulties in overt object recognition but nevertheless retained the ability to grasp objects. She could adjust her grasp aperture (how much she opened her hand, moving her thumb and forefinger apart) to object size. Strikingly, she was able to put a card into a slot oriented at various different angles, rotating her hand appropriately and early in the movement, well in advance of reaching the slot. Yet D.F. was unable to verbally describe the slot's orientation and could not adjust the orientation of her hand or of the card in order to report the slot's orientation. Goodale and Milner speculated that this dissociation arose from a breakdown of conscious ventral stream processing while unconscious dorsal stream processing remained intact.

It is noteworthy that this interpretation rests in part on a null result: that D.F.'s action system is unimpaired. Yet later research has shown that her visuomotor skills, including object grasping, are far from fully intact. She also has some residual conscious access to visual information, a result that is inconsistent with the claim of the two-streams hypothesis that only processing in the ventral pathway (damaged in D.F.) reaches visual awareness.[30]

A substantial number of studies on normal individuals have sought support for the two-route claim by comparing the influence of visual illusions on reaching versus perception. The Ebbinghaus illusion, shown in figure 2.5, is a good example. The figure comprises two equal diameter circles, one inside an annulus of small circles and the other inside an annulus of large circles. Typically these appear to differ in size. The conscious percept of size differences is assumed to reflect processing in the ventral pathway. If participants are asked to reach for one or the other of the target circles and to form their hand into an appropriate grasp and if grasp aperture turns out to be identical for the two circles, then this might suggest that the dorsal pathway and its connection to the motor system uses a quite distinct representation of object size than the ventral route. In particular, the conscious representation of the two circles as being of different sizes would not then be conveyed to the

Figure 2.5
The Ebbinghaus illusion. A circle surrounded by smaller circles (left) appears larger, while one of equal size but surrounded by larger circles (right) appears smaller. In a perception version, participants are asked to judge the relative size of the inner circles, a task assumed to depend on conscious ventral stream processing. In the action version, participants are asked to reach toward the inner circles as if they were aiming to grasp them, and the aperture of their hand opening is measured via infrared diodes attached to the finger and thumb. This version is assumed to involve unconscious dorsal stream processing and to be immune to the size illusion.

dorsal pathway, whose representations might hence plausibly be taken to be outside perceptual awareness.

Despite a number of reports of exactly this form of dissociation between perception and action, careful studies equating extraneous details of the perceptual and reaching tasks have found clear evidence that both systems are sensitive to perceptual illusions.[31] A team led by Goodale was the first to study this issue and reported that the illusion produced no effect on maximum grip aperture despite the fact that perceptual judgments were prone to the illusion.[32] However, other researchers conjectured that this dissociation was a result of a small but significant task difference between perception and action.[33] Whereas the perceptual judgments were based on a comparison between two circles presented side by side, each surrounded by the illusion-inducing circles, grasping responses were directed only at one of the target circles. When they avoided this procedural difference by having participants either grasp a single target circle or match its size by adjusting the diameter of an isolated reference circle, equivalent levels of illusion influence were obtained on both responses.

Debate concerning the interpretation of these and similar studies continues, but it seems fair to conclude that the strong form of the dorsal/ventral pathway hypothesis, and in particular the claim that processing of object information in the dorsal pathway is unconscious, has not been solidly established.[34]

Blindsight

At face value, blindsight is one of the most extraordinary neurological conditions. Individuals with this condition report being experientially blind in a part of their visual field (called a scotoma) yet are able to make a variety of discriminations about stimuli presented in that part of the field. The famous patient G.Y., for example, describes himself as having lost all the vision on his right, but he is able to judge (sometimes with high accuracy) whether a briefly presented object that he claims not to see is an X or an O or an angry or happy face.

Blindsight results from damage to primary visual cortex—G.Y. was injured in a traffic accident at the age of eight—and since external space is represented retinotopically (meaning that the relative organization of objects in the world is matched to their organization on both the retina and in parts of the brain) in primary visual cortex, there is a tight coupling between the location of the cortical damage and the location of the scotoma. Successful discrimination of location, movement, form, color, and so on, as well as overt actions such as pointing, have been reported in blindsight, and it has been proposed that these behaviors must be based on unconscious representations as blindsight patients deny visual consciousness regarding stimuli falling within their scotomata.[35]

For almost as long as blindsight has been investigated, the possibility that the condition is simply degraded (near-threshold) normal vision has been hotly debated.[36] This idea proposes that blindsight is conceptually similar to the state all people are in when they make judgments about barely perceptible stimuli. It is possible that residual visual discriminations with near-threshold stimuli are accompanied by weak, but reliable, levels of visual awareness. In fact, individuals with blindsight often report forms of visual experience. Alan Cowey noted in regard of D.B., the patient whose performance led to the coining of the term *blindsight*, that "there is still no explanation . . . for the revelation nearly 30 years after his operation, that he experiences visual after-images when a visual stimulus is turned off. . . . How ironic if the discovery of blindsight proves to be based on a patient who does not possess it!"[37] G.Y. also frequently reports awareness of stimuli in his "blind" field.

The other crucial component of the degraded normal vision hypothesis is that individuals with blindsight adopt an extremely conservative reporting bias. We all know somebody who offers an answer to every quiz question but whose accuracy does not match his confidence. We also know someone else who offers an answer, more cautiously, only when she is sure of being correct. These two people might have exactly the same underlying factual knowledge; where they differ is in terms of their willingness to offer an answer. The term *reporting bias* captures this difference, highlighting the fact that some people may have a liberal bias and others a more conservative one. From this perspective, there is little doubt that individuals with blindsight respond very conservatively when directly asked to report what they see; if in doubt, they report "not seeing." But this means that they might "see" more than they claim they do. As with other examples from neuropsychology, much of the evidence surrounding blindsight can be plausibly explained without recourse to unconscious influences.

The suggestion that blindsight is similar to normal vision under extremely degraded viewing conditions (though perhaps with an additional overlay of conservative reporting) raises the obvious question of its relationship to the highly charged topic of subliminal perception. The history of this phenomenon, in which our behavior is influenced by very brief and unnoticed stimuli, stretches back at least as far as a study carried out by the market researcher James Vicary in 1957 but later shown to be at best a hoax and at worst fraudulent. Vicary claimed to have secretly flashed the very brief messages "Drink Coca Cola" and "Eat popcorn" to many thousands of moviegoers and, as a result, increased their purchases of popcorn by over 50 percent and of Coca Cola by around 20 percent. When interviewed, the movie theater manager claimed that no such experiment had ever taken place, but by then it was too late: the idea that subliminal advertisements can boost sales was born. A wave of concern swept through the American public, and the practice was rapidly banned.

Intermittently, it continued to provoke concern, however. In 1990 the rock band Judas Priest was taken to court by the parents of two young men who had formed a suicide pact. It was alleged in the court case that the band members had inserted subliminal messages such as "Try suicide" in one of their records. A band member later commented that if they had wanted to include such subliminal communications, their message would have been to buy more of the band's records. The case was dismissed.[38]

In the years since then, reliable evidence that subliminal images or messages can influence meaningful behaviors over a timescale of more than a few seconds has been elusive.[39] In the laboratory, in contrast, many dozens of experiments have found more interesting findings. In a typical procedure, participants have to make a simple decision such as whether a face shown in full view has a happy or angry expression or whether a digit is less or greater than 5. A brief time before this "target" stimulus is presented, another stimulus (the "prime') is flashed very briefly, for a few thousandths of a second. This prime is relevant to the target stimulus: for instance, when the targets are happy or angry faces, the prime may itself be a happy or angry face. Thus, the procedure allows us to ask whether a very briefly (perhaps subliminally) presented face or word can affect responding to another stimulus.

What such experiments typically find is two things. First, the prime does indeed affect responding to the target, often by speeding up or slowing the decision. So we are quicker to judge that a face is happy if it is preceded by a briefly flashed happy face and slower if it is preceded by an angry face, suggesting that some information about the prime makes its way into the brain to slightly facilitate or impair processing of the target. Second, this effect occurs even when the participant reports not seeing the prime stimulus. If the participant is asked to state, at the end of each trial, whether she saw the prime ("seen") or not ("unseen"), results show that responding to the target is affected even by primes identified by the individual as subjectively unseen and invisible.

Just as with blindsight, the weak point in these experiments is the claim that the research participants are indeed unaware of the crucial stimuli that they describe as unseen. The concern is that these apparent subliminal effects simply reflect near-threshold conscious perception and reporting bias. The participant may just have a preference to label very fleeting images as unseen. And a raft of research has shown that dichotomous measures that ask the individual to report (yes/no) whether a stimulus is visible systematically underestimate the extent of visual awareness (regardless of response bias). When individuals (both normal and with blindsight) are given the opportunity to report the clarity of their perceptual experience using a range of categories such as "no experience," "brief glimpse," "almost clear experience, and "clear experience," stronger correlations are observed between awareness and choices than is the case when awareness is measured with binary responses.[40]

Valid Assessments of Awareness

This discussion about subliminal perception makes it clear that we always need to keep a fundamental question in mind when evaluating claims about the unconscious: Has the extent of awareness been thoroughly and exhaustively measured? In terms of theoretical significance, there is a world of difference between truly unconscious perception on the one hand and near-threshold but conscious perception on the other. Even marginal levels of awareness may be sufficient to explain behavior, without any need for recourse to the notion of unconscious perception. If our method for assessing awareness is poor, we may end up mistaking marginal awareness for a true lack of awareness. Indeed a persistent theme in the history of research on the unconscious, as noted earlier in this chapter, is repeated cycles in which apparent unconscious effects fail to hold up when more thorough and precise measures of awareness are available.

The criteria that need to be met by adequate awareness measures have been the subject of extensive debate in psychology for decades.[41] Here, we take a fairly simple approach and assume that the more reliable, relevant, immediate, and sensitive an awareness assessment is, the less likely it is to be distorted by insensitivity or bias or error. Table 2.1 provides brief explanations of these criteria (in the final chapter we return to the concept of validity).

An assessment will tend not to be *reliable* if it is influenced by factors, such as experimental demands or social desirability, that do not influence the behavioral measure. Prejudiced people may be fully aware of their prejudices but unwilling to freely admit them for fear of social disapproval. *Relevance* refers to the requirement that the assessment should not ask about information that is irrelevant to the behavior. The key issue is to what extent people are unaware of the information triggering their decision at the point of choice (*proximal* cues), as compared to information in the past (*distal* cues) that might have caused the current information (thoughts) to be present at the point of choice. Consider a situation in which some distal cue (as a child, you watched a nature documentary about the Galapagos Islands)

Table 2.1
Criteria for adequate assessments of awareness

Criterion	Explanation
Reliability	Assessments should be unaffected by factors that do not influence the behavioral measure (e.g., experimental demands, social desirability)
Relevance	Assessments should target only information relevant to the behavior
Immediacy	Assessments should be made concurrently or as soon after the behavior as possible
Sensitivity	Assessment should be made under optimal retrieval conditions (e.g., same cues are provided for measuring awareness as for eliciting behavior)
Psychometric soundness	Assessments should have sound properties (e.g., yielding the same measurement when repeated)

caused a proximal cue (your current positive attitude to those islands), which in turn influences a current decision (to book a trip to go there). Although you might be unaware of the distal influence on either your current attitude or your decision, you might be perfectly able to justify your decision in terms of your proximal attitude. Under such circumstances, it is plainly inappropriate, and a violation of the relevance criterion, to claim that the decision is influenced by an unconscious factor.

Immediacy refers to the fact that assessments will invariably be prone to forgetting or interference if they are taken some time after the behavior itself is evoked. Ideally, awareness should be monitored concurrently with the decision-making behavior itself, so long as such a measurement does not influence the behavior (in which case, awareness will need to be measured as soon after the behavior as possible). *Sensitivity* requires that the assessment be made under optimal retrieval conditions, such that the same cues are provided for measuring awareness as for eliciting the behavior of interest. In addition, awareness assessments need to be *psychometrically sound*, a requirement that is more likely to be achieved if continuous scales or scales with multiple response options (such as Likert scales) are used, which encourage the reporting of low confidence or partial knowledge, rather than binary ratings. Another aspect of soundness is that the assessment should, for instance, yield the same measurement when repeated, the property known as test-retest reliability.

Many of these criteria are not met by studies claiming to show unconscious influences on behavior. Although it may seem obvious that, for instance, an awareness assessment must target information that is relevant to the decision (*relevance*), experimental tasks sometimes prompt violations of the criterion.

There are, in summary, a number of important criteria that must be met in the design of an adequate awareness assessment. Although these requirements are extensive, it is important to note that the criteria are not unrealistic or unattainable. Many of the studies we describe in this book took considerable pains to deal with these issues of awareness measurement by measuring awareness concurrently with performance or via multiple convergent questions that are reliable, relevant, immediate, sensitive, and psychometrically sound.

* * *

The extensive literature on perceptual-motor skills provides many examples of the possible intrusion of unconscious influences on decision making, and it is undoubtedly part of our folk conception of such skills that they are influenced in this way. The amount of space we have devoted to them is tiny compared to the influence and scale of research on these topics. Yet careful examination of particular skills such as ball catching yields surprisingly little evidence that behavior is guided unconsciously. Under carefully controlled conditions, people seem able to report the heuristics they employ to catch balls, for instance. Libet's and Wegner's claims that conscious intentions do not cause our actions have been challenged on both philosophical and empirical grounds, and claims for the existence of unconscious routes to action based on neuropsychological conditions such as agnosia and blindsight have proven equally controversial.

3 The Ripples of Activation

Meet Donald: Once Donald makes up his mind to do something, it is as good as done no matter how long it might take or how difficult the going gets. Only rarely does he change his mind, even when it might well have been better if he had. What kind of person does Donald sound like to you? Persistent perhaps, or simply stubborn? People given this description above are in fact equally likely to endorse both traits as characteristic of Donald. But if someone has previously been shown the word *persistent* or *stubborn* in an apparently unrelated context, then the word they saw influences their judgment. What is going on here?

Welcome to the world of priming. The Donald study, published by Columbia professor Tory Higgins and colleagues in 1977, was the first to use the so-called unrelated-studies paradigm to investigate carryover effects: the idea that material (usually words) encountered in one context can carry over and affect our behavior in another, unrelated, context. But not only that: this carryover effect is said to occur involuntarily and outside of awareness.[1]

In the Donald study, the presentation of the *prime* word (*stubborn* or *persistent*) was incidental, buried within the demands of another task designed to distract participants from the main goal of the experiment. Half the participants were primed with positive trait words (for example, *persistent*), while the other half saw a negative synonym (for example, *stubborn*). After being exposed to the primes, participants went on to read the description of Donald and then provide a single word to describe his personality. Participants who had been incidentally exposed to *stubborn* were more likely to use a negative word; those given *persistent* chose more positive personality traits. This priming effect on personality assessment was enduring, being detectable in follow-up research even when the interval between the prime and the later evaluation of Donald was twenty-four hours.[2]

Crucially, this priming effect was not due simply to people choosing the word they had seen previously. Only in about half the cases did participants write down a word actually presented during the distraction task. On the other occasions the chosen words were synonyms, such as *determined*. This seemingly minor detail is important because it establishes the idea that broad positive or negative personality traits, and not just isolated words, can be primed incidentally.

From these relatively innocent beginnings, this unrelated-studies task has been adapted and extended with ever more surprising claims made about the kinds of behavior that can be influenced by simple primes. These studies are a natural progression if one accepts the iceberg view of the mind. Not only are decisions about ball catching and arm movements below the waterline, but also decisions that recruit what, on the face of it, would seem to require deliberative thought. However, as we saw in chapter 2, once one scrutinizes the assumptions of the ball-catching and the clock-face experiments, the evidence that such decisions arise from the murky depths of the unconscious begins to look rather shaky. Might the same be true for priming?

Priming Thoughts and Behavior?

> Disbelief is not an option. The results are not made up, nor are they statistical flukes. You have no choice but to accept that the major conclusions of these studies are true.

So wrote Daniel Kahneman in his best-selling book, *Thinking, Fast and Slow.*[3] The findings he was referring to are high-profile examples of how the unrelated studies paradigm has evolved over the past forty years. The mere fact that Kahneman had to advise his readers of the "need to believe" reflects the reality that at first glance, many of the findings he reviewed appear *un*believable. But not all priming effects require a leap of faith, so what is special about the ones Kahneman discusses? To answer that question, we first need to consider some research on what we might call basic priming effects.

Take a look at these two pairs of words:

BREAD NURSE

BUTTER BUTTER

The words on the left are clearly related: when we think about bread, we often think about butter. Thus, reading the word on the top line can act to prime our readiness to read the word on the bottom line. The pair on the right does not share this relationship: reading about a nurse does not make us think about butter. This difference in the relatedness of the words is reflected in the speed with which we identify the pairs as being words as opposed to nonwords. Specifically, people react much more quickly when asked to judge whether those in the left pair are both words than when asked to judge the right pair.[4]

This effect, known as semantic priming, is highly robust and replicable. As a tool, it forms the bedrock of an enormous amount of research in psycholinguistics, the study of how language is processed in the brain. It has been observed in many hundreds of experiments with thousands of participants. There are many varieties. Here's another: if you were asked to spell the word *sight*, you may be inclined to spell it differently depending on whether the question arose in the context of discussions about sensory modalities (*sight*), locations on the Internet (*site*), or references in a book (*cite*).[5] One simple explanation of these effects relies on the idea that words and concepts are arranged in our memory as nodes in an interconnected network. When we access one of these nodes, like *butter*, nearby areas of the network are also activated, thereby making it easier to retrieve related words, like *bread*. This notion of *spreading activation* throughout a network provides a powerful way to think about how our behavior is affected by the context in which we encounter information.

An even simpler example of priming can be illustrated using the pictures shown in figures 3.1 and 3.2 below and over the page (don't look at the second one yet!). Figure 3.1 contains a hidden image, and if you've seen the picture before, you will be able to see the image immediately. If you haven't, then it may take some time, but once you turn the page and look at the picture in which the image is clearly outlined, you'll never be able to unsee it!

Successful identification of the image in the picture can induce one-shot learning (priming) and affect perception of the same image years later.[6] This is an example of repetition priming, in which some response to the second presentation of a word, picture, or other item is altered as a result of an earlier presentation of the same item, often a long time previously. In terms of the network idea, it can be thought of as access to a particular node being strengthened as a result of repeated activation.[7]

A hallmark of these basic priming effects (repetition, semantic) is that they appear to be quite specific. The image of the dalmatian primes the

Figure 3.1
Can you see an image hidden in this picture?

same image, but it does not facilitate recognition of hidden images in other pictures. Reading *bread* primes *butter*, and probably *jam*, *knife*, *plate*, and other related concepts in a semantic network—but exactly how far does this activation spread? Kahneman argued that "mapping of these ripples" of activation through a network of associated thoughts is "now one of the most exciting pursuits in psychological research."[8]

It is important to ask why this pursuit is exciting. Is it exciting because it takes us on a journey that defies our commonsense conceptions of why we do the things we do? Similar to the ball catching and the clock face cases, there is something both unsettling but ultimately seductive and appealing about the notion that our behavior is being guided by mysterious forces that operate below the limen of consciousness. But this temptation to believe and feel as though we've understood something new or have an explanation of our decisions simply because we've ascribed them to a black box (or the murky depths of the iceberg) needs to be resisted. Being swept up in the pursuit of exciting, surprising, or sexy results is a root cause of the problems facing the science of the mind. So let's take a step back and look more closely at exactly how exciting these ripples of activation might be.

Pebbles or Boulders?

Try to rearrange the following words into a sentence (use four of the five words):

bingo / thirsty / plays / today / she

After a little thought it is easy to come up with, "She plays bingo today." What does that sentence make you think of? Does it bring a particular person to mind? What might that person be like? Is it more likely to be an old person or a young person? Bear that person in mind as you read on.

In essence, the current debate about priming effects boils down to two key issues: whether primes are more akin to throwing a pebble or a boulder in a pond and whether primes automatically trigger mental processes. Do the ripples spread without the involvement of any conscious, intentional processes?

Broadly speaking the pebble-in-a-pond view accords with that of many cognitive psychologists (those interested in theories and models of individuals' information processing). *Bread* primes *butter* and *jam*—and maybe a few other words, but that is about as far as it goes. The ripples tend to be weak, fleeting, and confined. Many social psychologists, however, adopt the boulder-in-a-pond view. The ripples don't just stay in the pond; the water splashes out over the banks and affects a whole range of other behaviors. For example, one prominent theory proposes that exposing people to words related to the concept of hostility (*hit, punch*) could lead them not only to be faster to identify the word *gun* but also to perceive another individual as more hostile (similar to the Donald experiment), behave in a more hostile manner themselves and become motivated to seek out an opportunity to aggress against some other person.[9]

To many of us, such claims are troubling. They paint the picture of a person who does not know their own mind, who abdicates responsibility for action to the vagaries of the current situation, whose behaviors are outside conscious control—and not just mundane or trivial behavior but consequential actions for both themselves and others. The number and variety of these behavior priming effects is truly astonishing. Some of the more outlandish claims include that we become more intelligent if we think about professors rather than soccer hooligans, that we think differently about our emotional closeness to our family members after graphing a pair of points

Figure 3.2
With the dalmatian now highlighted you will never be able to unsee it when you look back at figure 3.1.

close or far apart on paper, and that holding a hot cup of coffee leads us to evaluate a stranger more positively![10] Given their widespread implications for our understanding of human behavior, such claims need to be scrutinized. It would take too long to review every one of these "exciting" findings in agonizing detail, so we will focus on a couple of standout results that have been subjected to the required level of scrutiny and have, as we'll see, been found wanting.[11]

Walking Slowly

Let's return to the woman playing bingo. Was she old or young in your mind's eye? If you are like participants in what has become a classic study in social psychology, then imagining a person playing bingo should lead to thoughts of old people—and not only thoughts. The concept of old age then permeates—the ripples spread out—to affect other aspects of your behavior—specifically, how fast you walk. In the ingenious study, conducted at Yale in the 1990s, undergraduate students were given a series of scrambled sentences like the bingo-playing woman, all aimed to prime stereotypes of the elderly.[12] An additional group of students were given a different control set of sentences, equally difficult to unscramble but containing no age-related words. After completing the sentence task, participants were thanked for doing the experiment and directed to the elevator, down the hall from the lab room. And here's the clever part: another experimenter seated at the other end of the hall (apparently waiting for an appointment with a professor) surreptitiously timed how long it took for the participant to walk from the lab to a predetermined spot in front of the elevators approximately 10 meters away.

The striking finding was that participants primed with elderly stereotypes walked more slowly than those given the control sentences. The difference in time was not huge—on average about 1 second slower—but it was statistically reliable and replicated in a follow-up experiment. Moreover, none of the participants claimed to be aware of the relevance of the words in the scrambled sentences to the elderly stereotype, and none of them thought that the words could have influenced their subsequent behavior. This "walking study" has become, for many people, the poster child for the behavior-priming field. It was the first high-profile result to show that the unrelated-studies paradigm could be extended beyond concepts to influence actual physical behavior and outcomes. It is one of the studies that Kahneman suggested we must believe. As befits a famous example, however, it has attracted controversy and skepticism.

In part, this skepticism is driven by disagreement about the strength and extent of the ripples of activation. The authors of the walking study attribute the effect to the prime (old age) automatically activating a stereotypi cal trait (slowness), which mediates the walking speed (a behavior). But this strong boulder in a pond view is at odds with many basic priming effects. Indeed, a signature of those effects is that participants need to attend to the prime, and the prime itself needs to be strong and salient. You also typically need to be aware of the relation between the prime and the target of that prime.[13]

Armed with these questions and a few others, a team of Belgian researchers led by Stéphane Doyen and Axel Cleeremans set out to reexamine the walking study.[14] What they found was intriguing and suggests a need to pause for thought. The Belgian team had three main motivations. First was to see if they could replicate the study. Could they get the same difference in walking speed if they tried to run the experiment in a new location with new participants? The second was the conceptual ideas just discussed: the walking speed result seems surprising if you subscribe to the pebble-in-a pond view of activation. Third was the more mundane issue of possible problems with the methods used in the original study, specifically, whether manual timing via a stopwatch was accurate enough to measure the relatively small differences in walking speed.

The first new experiment examined the timing issue. The basic setup of the experiment was the same as the original: participants came to a lab, completed the scrambled sentence task either with or without the elderly prime words, and were then directed down the hall to a second location. The innovation was the use of infrared sensors located 10 meters apart along the hallway. Crossing the beam of these sensors triggered the timer, allowing the researchers to compute the walking speed of each participant. Using the sensors avoided any human error induced by preemptive or laggard pressing of the stopwatch buttons.

What happened? Not much. Regardless of whether participants were primed with the elderly stereotype or not, they took about six and half seconds to walk the 10 meters down the corridor. This was despite the fact that the new study had four times as many participants as the original Yale study. This larger sample size is important because, all else equal, having more participants, and thus more data, should make it easier to find evidence for an effect if one exists. So far, so bad for the boulder-in-a-pond view. Similar large-sample attempts to replicate the Donald personality assessment experiments with which we began this chapter have been equally unsuccessful.[15]

The second of the Belgian team's new experiments added another fascinating twist. They speculated that the original finding of difference in walking speed may have been due in part to what is known as an experimenter-expectancy effect. Put simply, if the experimenters (the people administering the tasks to the participants) knew in advance that a participant was given the elderly prime, they might have interacted with the participant in ways that induced stereotypical (slower) behavior. These changes in interaction could be conscious or unconscious on the experimenter's part; it does not really matter. What is crucial is whether participants somehow adapt to the experimenter's behavior and expectations and then walk more slowly—a kind of self-fulfilling prophecy.

To test this intriguing possibility the researchers manipulated the experimenters' expectations about the effect of the prime on participants' behavior.

Ten experimenters were recruited, half of whom were told to expect that the prime would decrease participants' walking speed and half of whom were told that the prime would lead to participants walking faster. Crucially, all experimenters knew whether a participant was in the prime or no-prime condition. The rest of the experiment was the same, with one additional feature: the experimenters were given stopwatches. This allowed the research team to compare the "subjective" timings from the stopwatches with the "objective" timings from the infrared sensors.

What happened this time? A lot. The subjective timings—those made on the stopwatches—revealed clear evidence of an experimenter-expectancy effect. Experimenters who were told that the prime should reduce walking speed did indeed record primed participants as walking almost a second slower over the 10 meters than participants who were not primed. However, experimenters expecting fast walkers recorded primed participants as walking faster down the corridor than those who had not been primed. Remember that the actual prime was still only for the elderly stereotype (slowness), so any increase in walking speed must either have been genuine and due to the influence of the experimenters' "fast" interaction with the participant, or "all in the mind" of the experimenter and due to being trigger-happy with the stopwatch.

The objective timings reveal which of these possibilities is more likely. When measured by the infrared sensors, the difference between prime and no-prime in the fast-experimenter condition completely disappeared: both groups took about six seconds on average. So the difference in the stopwatch timing appeared to be down to preemptive button pressing by the experimenters because they expected the participant to walk more quickly. Even more interesting, the objective timings for the "slow experimenters" still revealed a small but reliable effect of the prime. Those primed with the elderly stereotype did walk more slowly!

The message from this rather complex set of findings is that behavior can be primed—the ripples can spread beyond the edge of the pond—but it seems that such strong effects can occur only when the experimenter knows what the participant "should" do and somehow communicates these expectations.[16] A pure prime in the absence of this favorable experimenter context does not appear to be sufficient.

One last important finding is that the Belgian team interrogated their participants' awareness of the primed category (elderly) and its possible impact on their behavior. They found that most participants realized what the category was, and many who actually did walk more slowly seemed to be aware that they had slowed down. This kind of verbal report evidence—people explaining their behavior after the fact—is often ignored or underweighted in studies of purportedly unconscious influences on behavior. In fact, timely interrogation of information that is relevant to the observed behavior often reveals a close alignment between awareness and behavior (recall the criteria for proper awareness measures listed in table 2.1).[17]

But what if experimenters try even harder to disguise the link between a prime and the subsequent behavior? Then perhaps these subtle influences can emerge. Maybe.

Smiling through Your Teeth

Grab a pen and place the blunt end in your mouth. First, try holding the pen with your teeth so that the pointed end faces down. What kind of a face are you pulling? It should be something akin to a smile—teeth bared and mouth stretched and curved up. Now move the pen so that you are holding it with your lips—what is your expression now? It should be more like a pout, with the lips pursed and the mouth drawn down (see figure 3.3). Do you think your mood, or the way you felt changed as a result of these two

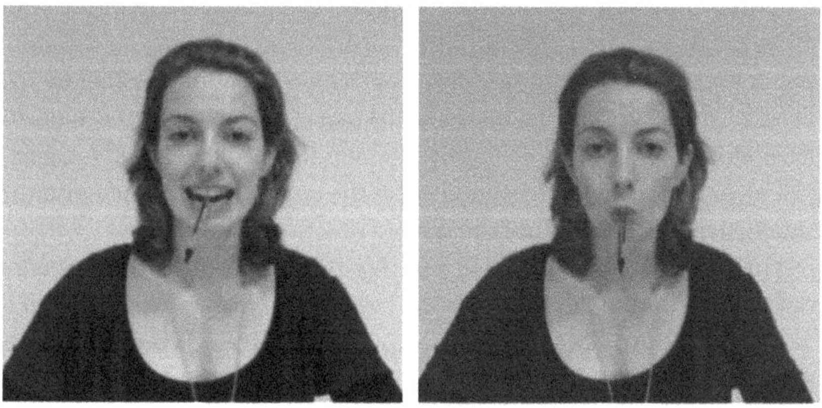

Figure 3.3
Does holding a pen in your mouth with your teeth bared (left) make you happier than with your lips pursed (right). (Spoiler alert: not really.) Figure available at http://tinyurl.com/zm7p917 under CC license https://creativecommons.org/licenses/by/2.0/.

pen-holding positions? This might seem like a bizarre question, but according to proponents of the facial feedback hypothesis, facial expressions can influence people's affective responses even when the expression did not result from an emotional experience.

This is not a new idea. In his classic study of emotions, Darwin proposed that a freely expressed emotion will be intensified by outward signs of that emotion (a smile, for instance), whereas one that is repressed will be softened. The question posed by the team of German researchers who conducted the pen study was whether a person needs to be aware of their own facial expression for this strengthening and attenuation of emotion to occur. They concluded that awareness was not necessary.[18]

The experiment was simple. Participants rated how funny and amusing they found a set of humorous cartoons (Gary Larson's *The Far Side*). Half of the participants did the rating while holding a pen in the "smile" position and the other half made the "pout" with the pen. Importantly, participants were never told to try to make a smile or a pout. Indeed, the researchers took pains to conceal this aim by providing a cover story about how the study was "to do with psychometric coordination" and that the researchers were "interested in people's ability to perform various tasks with parts of their body that they would not normally use for such tasks."

With this (distracting) cover story in mind, participants were then given the set of cartoons to rate. As predicted, participants holding the pen in their teeth rated the cartoons as funnier (an average of just over 5 on a 9-point scale ranging from 0 = not all funny to 9 = very funny) than those holding the pens with their lips (average rating of just over 4). A follow-up study found the same pattern for a question about the level of amusement elicited when looking at the cartoons (although interestingly, that second study did not replicate the difference on the funniness rating).

These results appear to provide good evidence that the intensity of felt emotions can be influenced via facial feedback even when people are not aware that they are smiling or pouting. They key difference between the pen study and most previous investigations of the facial feedback hypothesis is that in those previous studies, people were asked to smile or frown. This allows for an intentional influence of the expression on the emotion—"I am smiling so I should feel happier." The pen study claimed to effectively break this conscious intervention, thereby demonstrating a pure or direct motor-program effect of muscular activity on emotion.

This claim is similar in spirit to the basic premise of the unrelated studies paradigm. A feature of the environment that is not brought to our attention (our facial expression) nonetheless has a significant impact on subsequent behavior (rating of cartoons), all without us being aware of the influence or having any control over it. Despite this rather unsettling conclusion, the pen result is another example of the kind of study that Kahneman argued we simply have no choice but to believe. Or do we?

Just like the walking study, the pen study has attracted a lot of attention over the years. It has been highly cited and is commonly discussed in introductory psychology courses and textbooks. But until 2016, it had never been independently replicated. This changed when a new initiative (inspired by the "replication crisis" discussed in more detail in part II) was born. The Registered Replication Report (RRR), championed by researcher Dan Simons along with the US Association for Psychological Science, is a method for pooling the research efforts of lots of different labs to focus on one particular study. The simple idea is to provide an unbiased, objective, and transparent way to measure the reliability and size of an effect. Size here refers to statistical properties and in essence is a measure of whether we should care about or believe that the effect is "real."

The RRR for the pen study involved seventeen labs from all over the world and tested almost two thousand participants.[19] The setup of the study followed the original as closely as possible, and detailed protocol and video-based instructions were provided to all participating labs. Putting the instructions on video was vital for avoiding the potential experimenter-expectancy effects that can plague studies of this kind (as we saw with the walking experiments).

The data from all the labs were then pooled and subjected to a stringent predetermined analysis plan. Committing to this plan in advance protects against the kinds of data slicing and dicing and fishing expeditions that often lead to false positives (finding an effect when it isn't there). The key goal was to estimate a meta-analytic effect size—a fancy of way asking whether the effect is real if we look at all the data. (We'll read much more about these things later in the book.)

What happened? Recall that in the original study, smilers rated the cartoons around 5 on the funniness scale and pouters around 4—a difference of nearly a whole unit. This might not seem like much, but it was statistically significant and suggests quite a large effect. However, in the replication, which had just over twenty times as many participants, the mean difference between ratings made in the smile and pout conditions was only 0.03—essentially nothing. The authors of the RRR cited this as a "statistically compelling" failure to replicate the original pen study. It is important to stress, however, that there remains general support for the facial-feedback hypothesis but, crucially, only when people are aware that they are being asked to smile or frown. Once this conscious link between making the expression and experiencing the emotion is broken, there appears to be no feedback effect.

Reflecting on the failure to replicate, the author of the original study, Fritz Strack (who was not involved in the RRR), raised several objections about the way in which the replication had been conducted and how the data had been analyzed.[20] Probably his most bizarre claim was that the *Far Side* cartoons participants rated were no longer as funny as they had been in the 1990s when the original study was conducted! This critique was leveled despite the fact that the team conducting the replication took pains to obtain a set of *Far Side* cartoons that had been prerated by current students as appropriately funny for inclusion in the experiment. Many of Strack's other claims about apparent patterns in the data supporting the original conclusion could have been checked via the appropriate statistical analysis. Despite the data being freely available, Strack chose to leave his claims as unsupported speculations.[21]

A Mirage of Ripples

In one sense, the results of these failures to replicate are depressing. We thought we knew something fundamental about human behavior, yet deeper scrutiny suggests that we might be wrong. Yet in another sense, these findings are liberating and refreshing. Science is by nature incremental. We are building our house of knowledge brick by brick, and if that means that sometimes walls need to be knocked down or remodeled, then we should consider that progress.

More important, these attempts to tackle the surprising, sexy, and often counterintuitive findings head-on using the best available methodological and data-analytic techniques provide a tonic to the popular zeitgeist for easily led, irrational humans who are guided by their unconscious. It may not be quite as interesting to discover that thinking about old people doesn't make you walk slower or that holding a pen in your teeth doesn't make cartoons funnier—but science should not be about interesting; it should be about truth.

The walking slowly study and the pen study are just two of the ones for which Kahneman suggested that "disbelief is not an option." In the same chapter, he discussed several other surprising findings from a range of different "unconscious" priming experiments. The pattern across these experiments appears to lend weight to the idea that our behavior is influenced and controlled (in large part) by factors completely outside our awareness.

But this is a mirage. A systematic reanalysis of all the findings Kahneman cites as "not statistical flukes" suggests that they are indeed just that. Ulrich Schimmack, a Toronto-based psychologist, subjected the findings discussed by Kahneman to a replicability analysis. In essence, Schimmack tried to estimate how replicable the findings would be. As we've seen, the walking and pen study already do not seem watertight, and that was true too for the other priming studies. In summarizing the analysis, Schimmack wrote that readers "should disregard Kahneman's statement that "you have no choice but to accept that the major conclusions of these studies are true." Our analysis actually leads to the opposite conclusion: "You should not accept any of the conclusions of these studies as true."[22]

In a final twist in the tale, Kahneman responded to Schimmack's analysis by saying that he had not "unbelieved" the original studies he discussed and that implausibility is not sufficient to justify disbelief. He did, however, concede that we should be wary of relying on memorable—but not necessarily methodologically sound—studies as providing good evidence for scientific claims.[23] But for how long should we suspend disbelief? How many failures to replicate do we need to see before we start to unbelieve? As consumers of research on the science of the mind, it is important to know which of the barrage of findings out there we should be paying attention to. Implausibility may not be sufficient, but it should certainly raise concerns.[24]

We began this chapter by looking at some standard priming effects, including the semantic priming induced by reading "bread" before "butter" and the repetition priming of the dappled Dalmatian. In some sense, these effects appear rational, whereas the walking and pen studies seem distinctly irrational. If one were designing a system for the rapid decoding of letter strings, then it might make sense for it to be biased by what was perceived a few tens or hundreds of milliseconds previously. If one were designing a system for identifying hidden objects, it might make sense to allow it to access and be influenced by memories of similar objects seen in the past. But how can it be rational for judgments about our emotional closeness to our family members to be affected by the proximity of a pair of points we have connected on a sheet of paper or for our judgments of risk to be influenced by the activation of romantic thoughts?[25]

Perhaps neither (ir)rationality nor implausibility alone is sufficient for unbelieving, but in combination, they present a powerful case for reclaiming an admittedly less exciting, but probably more accurate, account of the science of the mind. But perhaps we are being too curmudgeonly. Surely there must be something to these claims that we can be nudged into making choices that we otherwise wouldn't, or influenced by seemingly irrelevant details of the environments in which our decisions are made. If we weren't, then why do we always end up coming back from the supermarket with several items we didn't know we needed (and often not the ones we did!). The next chapter shows that there are such impacts on our behavior that are all real and robust effects, but they do not rely on ripples of activation and have nothing whatsoever to do with unconscious influences.

4 The Leaking of Information

We have seen how the ripples of activation in the mind are typically fleeting and limited in their influence. Reading *nurse* will prime *doctor*, but unscrambling sentences about old people will not (necessarily) make you walk slower. Thus, in our quest to explain behavior and decision making, it appears that we do not need to rely on an intelligent, sophisticated unconscious mind that mysteriously takes control of our actions. However, we need to be careful here that we are not throwing out the proverbial baby. In many situations, it seems that very subtle features—such as the presentation of options on a page or the emphasis on one aspect of a product over another—can have large impacts on our decisions.

In this chapter, we look at these kinds of influences and ask how they operate. The main idea we pursue is the notion that information can "leak" from the way a question is asked, or a problem is framed, to the person making the decision or choice.[1] Such leakage is subtle but it turns out that we are extremely well attuned to these cues. Rather than providing evidence for influences from below the limen of awareness, these instances of information leakage show how well adapted we've become to consciously navigating our world.

10 Percent Fat or 90 Percent Fat Free?

Which do you prefer? The yogurt that declares itself to be "90 percent fat free" or the one that is "10 percent fat"? In terms of fat content, the two are clearly identical, but the former is likely to be much more appealing than the latter (no one likes to be reminded about how much fat they are eating). When you read these descriptions side by side, their mathematical equivalence is clear, but there is more to it than the math. Somebody (the advertiser, the product designer) chose to emphasize one aspect (90% fat free) over the other, and when we read the description, we know that they made that choice, and, presumably, we know why (because they want us to think the product is healthy and good for us). The choice of words plainly matters and implies a desired outcome or behavior—buy *this* yogurt, not the one next to it on the shelf. This communication between the "sender" and the "recipient" of the information is relatively subtle, but its influence can be significant.

The fat-free yogurt example is just one illustration of a broad class of phenomena typically categorized as *framing effects*. These effects occur when equivalent frames lead to different choices.[2] In this case, a single attribute—the fat content of the yogurt—is described in two different but equivalent ways. Moreover, the different descriptions have conflicting valences—one is good (90% fat free) and the other is bad (10% fat).

Crucially, in experiments examining the effect of these frames, the two frames are usually never presented to the same person side by side, or the game would be up—presumably, people would see that the two types of yogurt are equivalent.[3] But when the frames are presented in isolation, do people interpret them equivalently? While it might be true that they are logically or mathematically equivalent, as we will see, the *information* that the different frames impart may not be. Put simply, the choice of what proportion to describe conveys information in itself. The information leaks out of the description. In other words, these experiments show that people like the 90 percent fat-free yogurt more than the 10 percent fat, the 75 percent lean beef more than the 25 percent fat,[4] and indeed the medical treatment that promises 75 percent survival chances more than the one predicting a 25 percent mortality rate.[5]

Half-Full or Half-Empty?

Imagine there are two glasses on a table in front of you. One is full of water, the other empty. You are asked to pour water from one glass to the other and place a half-empty glass at the edge of the table. Once you have poured the water you will be faced with two glasses both with water at the halfway point, just like in figure 4.1. One of these glasses was initially full and one was initially empty. Which one do you pick to place at the edge of the desk?

Remember you were asked to place a half-empty glass, so either one would be correct in the sense of satisfying the request, but it turns out that the way

Figure 4.1
Are these glasses half full or half empty? Your answer will depend on whether the glass was previously full versus empty. (Photo credit: Zoila Newell.)

in which you are asked influences your choice of glass. When asked for the half-empty glass, almost 70 percent of people select the initially full glass. However, if the initial request asked for a half-*full* glass, then only 46 percent of people furnish the initially full glass.[6] Before we unpack what is happening here, let us look at a few more examples of this information leakage.

Imagine you have just flipped a fair, unbiased coin seven times and obtained the following outcomes (T = tails, H = heads):

T T T T H T H

You are then given a form like the one in figure 4.2 in which you can choose between two logically equivalent descriptions of the sequence.

What would you do? Given the observed sequence and the request, it would be correct to circle "heads" and "2," thus creating the statement "The coin came up heads two times out of seven." It would also be fine to circle "tails" and "5" to indicate that "the coin came up tails five times out of seven." Yet in experiments like this, three-quarters of people spontaneously choose the second frame: they describe the sequence in terms of the majority outcome.[7]

	circle one	*circle one*	
		0	
	heads	1	
		2	
The coin came up		3	out of 7 times
		4	
	tails	5	
		6	
		7	

Figure 4.2
How would you describe a sequence of coin flips that came up T T T T H T H? The vast majority of people do so by circling "tails" and "5" when given a form like this. Adapted from Shlomi Sher and Craig R. M. McKenzie, "Information Leakage from Logically Equivalent Frames." *Cognition 101*, no. 3 (2006): 467–494.

A final similar example should serve to cement the basic idea.[8] Imagine now that a sneaky experimenter has given you a loaded die but does not tell you it is loaded. The die is six-sided and has five black sides and one white side and is weighted so that it tends to fall on the single white side. You roll the die six times and then have to fill in a form like the one in figure 4.2—but with the words *die* and *black* and *white* instead of *coin, heads*, and *tails*. Let's assume the die came up: *black, white, black, black, white, white*. How would you choose to describe this sequence? Because the die has five black faces and only one white you might expect a sequence of mostly black outcomes—five times out of six it should be black (remember that you don't know the die is loaded). Perhaps to emphasize this deviation from expectation, people given this task tend to say: "The die came up *white* three times out of six." In contrast, if the die had been mostly white with only one black face and the same set of outcomes obtained, then the vast majority (83%) say that: "The die came up *black* three times out of six." In both cases, it appears that people describe the outcomes in terms of what has increased relative to an expected proportion.

These examples serve to illustrate that subtle linguistic cues influence our interpretation of situations in nuanced but important ways. They also demonstrate these influences in terms of both the way requests are made and interpreted, as well as the way we spontaneously choose to convey information. But what information, exactly, is leaking through?

A key part of the explanation seems to reside in the idea of a *reference point*: prior to flipping a coin, we have an expectation that we will see roughly equal numbers of heads and tails in a sequence of outcomes. Thus, our reference point value would be 50 percent. We can also assume that other people would share this same reference point. When we see a lot more tails than we might have expected, this increase relative to our reference point becomes salient and drives our description—hence, the preference for describing the outcomes in figure 4.2 as "five tails out of seven." The dice experiment provides even stronger support for this idea by showing that the description of the same set of outcomes is influenced by whether there was an expectation for outcomes of mostly one color or the other. That is, when the reference point is not 50 percent white but 17 percent (1/6), we choose to emphasize the increase relative to the reference point by framing the description in terms of white outcomes.

One question that arises here is why people tend to emphasize the increase relative to the reference point rather than the decrease. Why not describe the coin flips in terms of seeing fewer heads than expected? It is probably fair to say that we do not really have a satisfactory answer, but one possibility is that the salience of attributes in people's explanations of observations is determined by relative abundance rather than relative absence. A speaker's choice of what to emphasize is more influenced by attributes that there are a lot of in the object being described simply because lots of things are more salient than scarce things.[9]

Let's return to the glass-half-full example. This shows that people notice the departures from the reference point that are implied by a speaker's choice of frame. Let's assume that the initial state of the glass implies the reference point: after pouring, the initially empty glass ends up above the reference point, and the initially full glass ends up below it. If you are asked for a half-empty glass, then the reference-point hypothesis predicts that you are more likely to offer the initially full glass than the initially empty one. However, if you are asked for a half-full one, then you should choose the initially empty one. In simple terms, the reference point hypothesis suggests that people spontaneously think of half-empty glasses as ones that were previously full and half-full glasses as ones that were previously empty. The fact that both the speaker and the listener share this implied assumption is what leads to the observed results.

We can see the same process in action in the meat, yogurt, and medical treatment examples. According to the information leakage idea, a speaker's choice of which aspect of an attribute to emphasize is intentional, not accidental. Moreover, these statements do not appear magically inside our heads: we know that a speaker chose to select this particular way of framing the information. Thus, when we hear that a medical treatment has a 25 percent mortality rate, we are more likely to infer that this is an atypically high rate (because the speaker has chosen to emphasize it) than if the corresponding survival rate had been used. This leaked information from the frame then leads to a valence-consistent preference shift: I am less likely to choose the treatment when the mortality rate is emphasized than when the survival statistic is used.

The fundamental message of these studies is that attribute framing effects do not imply biased or irrational decision making. Quite the opposite: they suggest a remarkably sophisticated "conversation" between speakers and listeners (or at least experimenters and participants). The results are testament to the fact that the cognitive machinery we employ when interpreting our world is extremely sensitive to the cues that surround us. It's not hard to imagine that similar conversations take place in many other contexts, such as between a chef and a restaurant customer when the former decides to place a particular dish at the top or bottom of the menu and thereby subtly influences the likelihood of it being chosen.[10]

Are We Aware of Leaked Information?

The leakage studies provide clear evidence that mathematical equivalence is not the same as informational equivalence. But are we aware that these changes in frame are influencing our decisions? Craig McKenzie and Shlomi Sher, on whose work much of this chapter has focused, suggest that we are not. They write, "*Whatever inferences are involved are surely implicit—i.e., drawn below conscious awareness.*"[11] In other words, they see the sensitivity to the frames emerging from the depths of the iceberg—precisely those murky regions that we argue do not exist or at least do not play any role in determining our judgments and decisions. How do we reconcile these contrasting interpretations?

The following thought experiment might shed some light. Let's go back to the question about the lean or fat beef. Assume that you are inviting a friend to dinner and making their favorite lasagna dish with *80 percent lean* ground beef. Think about how you would rate the beef on various scales that asked about the lean/fat content, the greasiness, the quality, and the taste. Now imagine you were asked to imagine the same scenario but with the beef described as *20 percent fat* and you were given the same set of rating scales. Would your ratings change?

This is an example of what is called a within-subject design—in which both versions of a question are given to the same participant. Experimenters interested in the quirks and "irrationality" of judgments often avoid this kind of design, instead opting for between-subjects ones where different participants are given the different versions (one group gets the 20 percent fat wording and the other the 80 percent lean). The reason for preferring to compare different groups seems obvious: when the frames are transparent, people surely would not be tricked into giving different ratings, and so no (irrational) framing effect would be observed.

Maybe. However, if what is happening is that people are explicitly sensitive to the information that leaks from the frame—that is, they are making conscious inferences about the choice of frame—then we'd expect to see differences in ratings for the beef even when the descriptions are presented one after the other. Intriguingly, this is exactly what is found in a within-subject version of the lean/fat beef experiment.[12] The same participant gives higher ratings for quality, taste, and other characteristics when the beef is described as 80 percent lean than when it's described as 20 percent fat. Moreover, it does not appear to matter whether you ask the two questions immediately, one after the other, or wait a week or two between asking the first and the second. We might expect that delaying the second question would increase the change in ratings: perhaps people forget their previous answer or even that they had been asked before. Apparently not; whatever the timing, people appear to infer that the two statements imply different things about the beef and they change their ratings accordingly and to the same degree whether they were asked just now or last week.

This pattern of results raises the question: Why does the inference need to be implicit in the sense of being below the limen of conscious awareness? People see the information in the statement; they understand explicitly what the information implies, and they make an appropriate rating, judgment, or decision. Our claim is that this is the right way to view many types of influences on our behavior, not just framing effects.

Nudged or Inferred?

When an architect designs a building, many decisions are made about the placement of doors, windows, walls, and staircases that will affect the behavior of the building's inhabitants. People who design our choice environments play a similar role: choosing what aspect of a product to emphasize, the order in which to present information, and how to frame requests. Thus "choice architects," to use the term coined by Richard Thaler and Cass Sunstein, capitalize on the fact that a choice can never be presented in a neutral way: any way in which a choice is presented has the potential to influence—or nudge—how we choose.[13]

One of the best-known and most robust effects in the nudge playbook is the default effect. Put simply, people are more like to "choose" an option if it is presented as the default. The scare quotes here are intentional because one of the key aspects of the default effect is that a person is not necessarily making an active choice; it may be more of an acquiescence to a predetermined state of affairs. So does this pervasive and strong impact of defaults constitute evidence for an unconscious influence on behavior? Are we once again in a situation where our minds are being made up for us?

Consider the case of organ donation. What to do with functioning organs in the event of someone's death is a complex decision, emotionally laden and ethically charged. Many countries, in acknowledgment of this complexity, put policies in place either to mandate or encourage a particular course of action. Registering your intent to be a donor is perhaps the simpler option: this makes it clear that if you die, then your wish is for your organs to be used to help save others' lives. But what if you've never registered?

It turns out that different countries have different answers to this question. Some explicit-consent countries presume that no one is an organ donor unless they have registered. In other states by contrast, if an individual has not taken any action, then the default is to presume consent—that is, to presume people are donors unless they've registered not to be. In other words, some countries have an opt-out default and others have an opt-in. Does this difference in default types influence organ donation rates?

Eric Johnson and Daniel Goldstein examined this question and what they discovered was startling. Focusing on European countries, they found that opt-out nations had consent rates of over 90 percent, whereas opt-ins hovered between 5 and 20 percent.[14] This held true even when considering very similar countries. For example, although Austria and Germany are very close in terms of culture, socioeconomic status, and geographic location, only 12 percent of Germans were donors (in 2003) compared to 99.98 percent of Austrians. By simply changing the default option, the percentage of the population consenting to organ donation increased dramatically. Moreover, this increase in consent translated into an actual rise in the number of organs donated in many of the countries studied. And this is despite the fact that it was trivially easy for a German to register as a donor and for an Austrian to opt out. The act of opting in or out, despite its simplicity, seems to create enough friction to deter many people from doing so who presumably would if it were completely frictionless.

The ideas inherent in the information leakage account of behavior can again shed some light on the psychology of defaults. The whole premise of nudging and the choice environments that choice architects build for us is that no decision is made in a vacuum. As we have seen repeatedly, the frame matters: people are sensitive to subtle cues, which leads to a dialogue of inferences between those imparting the information and those receiving it. Defaults are no different.

To get a better sense of the logic, answer the following three questions:[15]

1. Are you willing to be an organ donor after your death?
2. In general, do you think people ought to be organ donors?
3. Imagine you can choose the organ donation policy for your country/state. Should it be one where the default is to be a donor or where the default is not being an organ donor?

Participants given these questions in a survey showed a clear relation between their own willingness to be a donor, whether they thought other people should be a donor, and what the government policy should be. Specifically, participants who selected donation as the default in the third question were more likely to be willing to be donors themselves and more likely to think other people should be donors, than those who selected the nondonation default. Thus, participants' own preferences predicted their chosen default.

In a follow-up experiment, one group of participants read about some policymakers who had decided to make organ donor the default and a second group read a similar statement in which policymakers had selected "not an organ donor." When then questioned about the beliefs and motives of the policymakers, participants given the donor scenario were more likely than their not-a-donor counterparts to infer that the policymakers would be personally willing to be donors and that they thought other people should be donors.

These simple, intuitive demonstrations show that default effects may occur in part because information leaks from policymakers' choices of available options. The beliefs and attitudes that policymakers have about a given issue like organ donation can be inferred by the public simply through the choice of the policy default. If a government preselects a donor default, it is implicitly recommending organ donation as a good thing for citizens to be doing. As a citizen, if I am aware of this policy, then I can explicitly infer that my government would like me to be a donor, and as a consequence, I probably would be more likely to stick with the default.[16]

It should not come as a surprise that we are sensitive to the different ways in which information is presented to us; we often employ default-setting in much more mundane circumstances. Imagine you are meeting a friend for lunch but the restaurant has not yet been decided. There are two options, Coco's or Barmilano, and you have a preference for Barmilano. If you were texting your friend and wanted to influence that person's choice, which of these two texts would you send?

a. I'll meet you at Barmilano at 1:00 p.m. If you'd rather meet at Coco's, then let me know.
b. I'll meet you at Coco's at 1:00 p.m. If you'd rather meet at Barmilano, let me know.

It seems obvious that the answer is a. By choosing that option, you are in effect setting the default; moreover, it seems likely that your friend will understand your preference for Barmilano.[17] Whether your friend goes along with the plan might depend on how strongly he or she prefers one restaurant over the other, and sometimes setting defaults can have the perverse backfiring effect of people choosing the nondefault option. However, current evidence suggests that people are often aware of why options are preselected, and even when the reasons for selecting them are explained (for example, "We know that in decision situations, people often stick with a choice option which is preselected for them. Therefore, we have preselected [option X]"), the information does not reduce people's willingness to follow defaults.[18]

Information leakage is not the whole story here, as default effects can also be linked to basic laziness on our part. If changing from the default introduces friction and requires effort—accessing forms, filling them in, sending them back—we are more likely to stick with the default whatever it is. But the impact of the inferences people make on the basis of selected defaults should not be underestimated. This is particularly true for situations in which people are uncertain about their own preferences and thus more likely to be guided by the (apparent) wisdom of others.[19]

Anchoring as Information Leakage

Ask yourself the following questions:

Do you think Aristotle was born before or after 1825 CE?

In what year do you think Aristotle was born?

Now imagine you'd been asked a different first question in which the year 1825 CE had been replaced by 25,000 BCE. In an experiment using these questions, people given the more recent comparator year (1825) estimated Aristotle's birth date to be about 140 BCE, but over 1000 BCE if they first judged whether he was born before or after 25,000 BCE. This is clearly rather odd behavior—Aristotle's birth year remains the same (384 BCE) regardless of the year to which it is being compared.

This simple example is an illustration of another kind of information that can leak into people's judgments: *anchors*. Tversky and Kahneman, in their incredibly influential 1974 paper in *Science*, described anchoring as one of three main heuristics (or biases) that people turn to when making judgments under uncertainty (the other two are *representativeness* and *availability*).[20] One interpretation of what underlies this apparently odd effect follows the same logic as the default effects: when people are uncertain of their preferences (or in this case, their inferences about a fact), they use information given, or leaked by, the provider to help them come up with an answer.[21]

The basic idea is that when you read "1825," you infer that the person asking the question has provided this reference date because it is useful for answering the question. So if you are unsure, you use that information as a guide (an anchor) and then recruit any other information you might have available until you reach a satisfactory response. The effect of the anchor is to limit how far you depart from this initial, presumably useful, starting point. Of course, if you know the answer, these implicit cues are irrelevant (imagine that the questions were about the birth date of a close family member). In fact, studies examining the impact of anchors show that the greater a person's expertise is,[22] or indeed just the more a person is incentivized to think carefully about the answer, the weaker the influence of the anchor.[23] Presumably both incentives and expertise act to reduce the uncertainty about the answer, leaving less room for the anchor to exert an effect on judgments.

Interpreted in this way, anchoring effects are entirely consistent with the view that people make judgments explicitly. This perspective also suggests that anchoring effects should be largest when people have the lowest confidence in their answers and smallest when they are sure they are correct. It also implies that the less one trusts a given anchor to be informative, the less one should rely on it.

Experiments examining the perceived plausibility of anchors lend some weight to this general conclusion, but some surprising effects persist. For example, even when a random number generator is used to produce an anchor value—something that should eliminate any kind of communicative intent on the part of the experimenter—anchoring effects are still observed, particularly in preferential judgments about how much people are willing to pay for consumer goods. However, the size of these effects still seems to be related to the distance between the anchor and the range of plausible prices that a given good might take. For example, if you are stating how much you would be willing to pay for an electric toothbrush, your price would be affected more by a randomly generated anchor that overlapped with your intuitive range of prices (for example, $25) than one that was too extreme (for example, $25,000).[24]

But the most compelling evidence that this influence is conscious is that people tend to view the anchor as providing useful information and are fully aware of using the anchor as a guide to their estimations. Indeed, a majority of people say that they want to see the anchor, even if they are told that it was generated by a roulette wheel! Moreover, they are influenced by the anchor only when they regard it as a good estimate of the quantity they are trying to judge: when they regard it as a poor estimate, no anchoring effect occurs.[25] The evidence is very clear in showing that people tend to regard the anchor as informative (even when it isn't), want to use it, and are aware that the anchor will affect their estimate.

One counter to this information leakage account, however, is the suggestion that anchors can affect our judgments *even when we are not aware of them*, or at least are unaware of them having any relevance to our judgment. If such findings obtained, it would create potential problems for the claim that anchoring effects result from the explicit integration of information. We cannot draw inferences about the usefulness or otherwise of a provided anchor if we have not perceived it or its relevance consciously. Do such automatic or implicit anchoring effects exist?

A simple study claims to find such evidence. Participants are shown a photograph of a restaurant and asked how much they would be willing to pay for a meal at this restaurant. Willingness-to-pay amounts were claimed to be larger when the restaurant was called Studio 97 than when it was called Studio 17, as if the number in the name had somehow primed them to think about and become willing to spend larger monetary amounts. Whatever the number means in the context of Studio 97, it presumably has nothing to do with money, or else the owners of Bistro One in Sydney ought to contemplate a name change if their ambition is to nudge their customers into spending as much as possible! Thankfully, restaurant goers need not be too concerned about being unconsciously parted from their money as a result of a cleverly chosen name. This and other incidental anchoring findings were not replicated in much larger follow-up studies that we and others carried out.[26]

In summary, it appears that when people are uncertain of an answer, are aware of an anchor, and trust that the provider is following conversational norms, anchors can have a very large effect on people's judgments. In contrast, when people have expertise, are incentivized, or have reason to disbelieve the relevance of an anchor, the impacts are smaller.

Causal Field Perspective

The picture that emerges from these studies of information leakage, defaults, and anchoring is one of sensitivity to subtle cues, which leads to a dialogue of inferences between those imparting the information and those receiving it. Should we then conclude that we are always aware of (all of) the information that leads us to make a particular choice?

Consider an experimental psychologist who wants to investigate the impact of mood lighting on diners in a restaurant. The psychologist wants to know if lower—and thus perhaps more romantic—levels of lighting lead to romantic relationships. To test this hypothesis, she invites different couples on blind dates to the same restaurant and ensures on alternate evenings that the lights are either dimmed or on full. She then records the number of couples who go on subsequent dates after their night out.[27]

If she found that more couples who dined in dimmed lighting met up again than those who dined in the bright lights, what would this mean? From the perspective of the psychologist, this would imply that dimmed lighting promotes future romantic engagements. Lighting level is what was different between the two conditions and thus stands out as the reason for the difference. It could of course be that bright lighting *discouraged* the second date, but no matter, the inference is that lighting has a causal influence on the likelihood of subsequent encounters.

But now imagine yourself in the shoes of one of the diners. What would you say if you were asked whether the lighting in the restaurant on your blind date influenced your decision to meet up again? Almost certainly you would deny any causal effect. You might go as far as being annoyed at such a preposterous suggestion and feel affronted that your affections could be swayed by something as trivial as restaurant lighting!

It is possible to argue that both you (the diner) and the psychologist are correct in your inferences and yet those inferences appear to imply different conclusions about whether we are aware of the factors that influence our behavior. The diners experienced just one of the lighting conditions—dimmed or bright—and thus have no way to attribute their subsequent dating behavior to a difference in lighting. Thus, they appear to be completely unaware of the causal influence that lighting has on the prospects for their love life. The psychologist, on the other hand, is in the privileged position of knowing about both conditions and draws a causal inference about the impact of the manipulated variable (light level) on behavior (dating).

The veracity of these conclusions hinges on the level of analysis that we seek. We are, of course, never aware of all of the factors that lead to a particular decision or behavior. When we type a sentence on a laptop, we are not aware of the exact combinations of neurons that have to fire to ensure that our fingers press the correct configuration of keys. Nevertheless, we are aware of what we want to write and our decision to carry out the necessary actions to do so. As we argue throughout this book, it makes no sense to ask whether we are conscious of the computational level of the brain's operation. What is important for our perspective is whether the factors an individual is aware of at the time a choice is made are sufficient to provide an adequate explanation of their behavior.

With this perspective in mind, let's go back to our couples in the restaurant and push the thought experiment a bit further. Now imagine that the psychologist sets things up as before with the dimmed and bright lighting on different evenings, but this time she blindfolds the couples on the dates. What is the prediction for this version of the experiment? Presumably because the blindfolded participants are insensitive to the different lighting conditions, we expect future romantic encounters to be equally likely regardless of the initial setting. Such a result would be instructive; it would tell us that the diners need to be aware in some sense of the lighting conditions on their first date for the improving effect on their love life to materialize.

The question then is, in what sense? If we went up to one of our original diners during the date and asked them about the lighting conditions, they would no doubt be able to tell us that the lights were dim or bright. They could probably also, if quizzed further, speculate on the effect those conditions were having on their general mood and enjoyment of the date. Perhaps the dim light provides a more relaxed pleasant ambience whereas the bright lights highlight the imperfections in our dining partner's complexion.

So, in the moment, the diner might well be aware of how environmental factors are influencing their current experience. The disconnect from awareness appears to come when they are asked afterward about their reasons for a different behavior—the decision to go on another date. At this point, the claim is that they would not immediately cite lighting as a reason for calling and arranging a new rendezvous. They would presumably say that they had enjoyed the company of their dining companion and wanted to see that person again—and these reasons would be a perfectly accurate justification for their current decision.

We met this idea in chapter 2 when discussing the criteria for adequate assessments of awareness. Recall that for an assessment to be *relevant*, it should ask only for information that is relevant to the decision at hand. That is, it should pertain to *proximal* cues—the information that is triggering a decision at the point of choice—rather than *distal* cues, or information from the past that might have caused the current information (thoughts) to be present. To adapt our trip choice example to the Galapagos from chapter 2 to the current situation, imagine yourself at the restaurant on the blind date trying to decide what to eat. In this situation, there might be some distal cue (your mother advised you as a child that garlic is a good source of antioxidants) causing a proximal cue (your current belief that garlic is healthy), which in turn influences a current decision (to select a garlic dish off the menu). You might be unaware of the distal influence on either your current belief or your decision but still be perfectly able to justify your decision in terms of your proximal belief. In this situation, it is simply incorrect to claim that the decision is influenced by an unconscious factor.[28] (It may be an unwise decision on a date, but that is a different matter).

The same interpretation applies to the lighting: the distal cue is the ambience affecting your assessment of the date and your partner (something you would have been aware of at the time), the proximal cue is your current belief that the partner is someone you'd like to see again, and the current decision is picking up the phone to arrange the next date.

A Leak or a Bias?

We have seen in this chapter how acutely sensitive we can be to the ways in which information is presented and how the decisions made by choice architects can influence the choices we make. We have argued that in large part, people are aware of these influences: information leaks from descriptions, frames, defaults, and people tend to pick up on it. Thus, again, we find little evidence for unconscious mental processes driving our choices and decisions. There is no question that there are subtleties to many of these influences, but in the moment, it appears that we can access the information that drives our behavior.[29] Scratching beneath the surface does not appear to provide any evidence of causal influence of what is purported to lie beneath the limen of conscious awareness.

In the next chapter, we scrutinize what this perspective means for the idea of unconscious biases: the claim that despite our best intentions, we are at times held hostage to prejudices that lie deep beneath our conscious awareness. As you might predict from the discussion to this point, determining if and how people are biased, unconsciously or otherwise, is much more complex—and interesting—than you might think.

5 Rethinking Unconscious Bias

The world is a very unequal place. In most Western companies and institutions, women and individuals from ethnic minorities are paid less and are less likely to be promoted than men. Of the top five hundred companies in the United States, only 41 had female chief executives in 2021.[1] Similar inequalities can be seen almost everywhere one looks. In Australia around forty people per 10,000 of the male population are in prison, but the figure for those of indigenous origin is a staggering 438.[2]

Bias refers to irrational preferences or tendencies that favor some choices over others.[3] Central to the concept is the idea that someone manifesting a bias tends to be impervious to contrary evidence or remediation. If the selection panel for hiring new members of a prestigious orchestra continues to prefer male over female musicians despite objective evidence that the candidates have equal musical ability, no one would disagree that this is an irrational and undesirable bias as well as being unfair.

Males and females are of course roughly equal in the population overall. But few would argue that because more men than women are incarcerated, the prison system is therefore biased against men. Rather the prison population reflects the overall difference in criminality rates between the sexes. The point is that a bias is a bias only if it runs counter to relevant hard evidence and impartiality. Gordon Allport, one of the pioneers of social psychology, coined the phrase "thinking ill of others without sufficient warrant" to capture the idea that misalignment with objective facts is central to concepts such as bias and prejudice.

Establishing sufficient or insufficient warrant can quickly become quite complex. The University of Oxford has been frequently castigated because so few of its students come from ethnic minorities. But is this a bias in the true sense of the term? About 17 percent of ethnic minority students are offered a place in contrast to about 26 percent of White applicants. Yes, it's a bias.

But hang on. Courses vary in how competitive they are, and it turns out that Black students are more likely than White ones to apply for the most competitive courses at Oxford (29 percent of Black students apply for medicine, a highly competitive subject, against only 7 percent of White applicants). So the fact that relatively fewer Black students are admitted to the university could be reflective of a bias, or instead could be a consequence of their tendency to apply for highly selective courses combined with a color-blind and unbiased admissions system.

But hang on. If we look at specific courses such as medicine, we find that highly able students (ones who got top grades in their final school exams) are much more likely to be offered a place if they're White (43 percent) than if they're Black (22 percent). So it is a bias after all.[4] The point is that establishing whether a preference is a bias or not—whether it is warranted—can be very difficult when the data on which the preference should legitimately be based are hard to quantify and unpack.

When it comes to examining the behavior of individuals rather than systems, establishing bias is just as difficult. Proving that people show unconscious or implicit biases, preferences that they're not even aware of, is harder still. Yet this notion has gained a great deal of traction in recent years, especially among employers, who spend vast sums annually on diversity training programs and interventions designed to eliminate prejudice in the workplace. It is now commonplace for individuals sitting on selection panels to be required to undergo unconscious bias training, offered with the entirely laudable aim of helping us to recognize bias and intolerance in ourselves and others and alerting us to conditions and circumstances in which we may be unintentionally susceptible to bias. Training programs designed to suppress unconscious bias use a range of techniques, such as encouraging us to deliberate rather than make snap judgments, reconsider reasons for or against our decisions, question cultural stereotypes, and monitor each other.[5] Yet the evidence that diversity training programs are effective at changing workplace behavior is murky at best,[6] and more worrying, scant regard has been given to the possibility that such programs could actually be harmful. There is research clearly showing, for instance, that some interventions designed to reduce prejudice—particularly ones that emphasize societal norms against prejudice—may backfire and in fact increase it.[7] It is therefore imperative that the case for unconscious bias is examined critically. This is the subject of this chapter.

A Kernel of Truth?

There can be little doubt that the human mind is capable of making extraordinarily subtle snap judgments and that our decisions are often based on only some of the relevant evidence. For example, a brief glimpse of around a tenth of a second is sufficient for us to decide whether a face is trustworthy or competent.[8] If such judgments are often inaccurate, it would clearly be irrational to base our decisions and preferences on such snap judgments. It is not a good policy to trust another person simply because something in her face makes us feel intuitively that she is trustworthy. But research findings illustrate convincingly that these snap judgments often tend to have some validity.

In one study, the cooperativeness of rural Senegalese men and women was established by asking them to play a four-person public goods game.[9] In this game, each player was initially given 200 grams of rice, which they could divide between a private and a public endowment. The rules of the game dictated that the total amount contributed to the public endowment by all four players was doubled and equally shared. Thus, players came away with an equal share of the public endowment plus whatever they held back as their private endowment. If each player allocated all of their rice to the private pool, they all ended the game with the same amount of rice as they started: 200 grams. If they all allocated all of their rice to the public endowment, each finished the game with 400 grams. However, it is possible to free-ride: a player who contributes nothing to the pool nonetheless benefits when the enlarged pool is shared. The game therefore measures the extent to which each player is willing to cooperate in a public endeavor to pool and multiply a good, and some of the Senegalese playing the game were found to be cooperative and others selfish. Can each person's cooperativeness be judged from his or her face? In the next phase of the experiment, pairs of photographs of these Senegalese players were shown to volunteers over two thousand miles away in Montpellier, France, who were asked to decide which member of each pair of faces was the more selfish person. Remarkably, the selfishness of male (though not female) players could be judged at levels better than pure guessing. A brief look at the face of a male from a different culture is sufficient to extract some valid information about that person's cooperativeness.

One area in which bias has, understandably, been extensively explored is in relation to gender and employment. In one controlled study, a job application was created, purportedly submitted by a recent science graduate who was applying for a laboratory manager position.[10] The application was assessed by a large number of science faculty at several US universities who rated it for how employable and competent they judged the applicant and what starting salary they would offer. Crucially, the applicant either had a female (Jennifer) or a male (John) name. Despite the fact that everything else in the application was identical, the female applicant was judged less competent and employable than the male one and offered a lower starting salary. Since the only difference between the applications was the individual's name and implied gender, the results seem to strongly imply gender bias in job evaluations and hiring decisions.

The problem with this conclusion is that discrimination and bias are not the same thing. Any employer who hires women at lower rates than men even when they have equivalent qualifications is guilty of discrimination in the legal sense of the term. It is contrary to employment law to base hiring decisions on gender. But this does not necessarily make it irrational to do so. There is much evidence showing that the lifetime productivity of female researchers is lower on average than that of males, not least because of career breaks to have children.[11] If only a minimal amount of evidence about a candidate is available, then gender is a valid signal of likely productivity. (Of course, one can legitimately question whether lifetime productivity is a fair criterion against which to measure success.) It is quite right that discrimination is illegal and that employers put policies in place to mitigate it, but that doesn't mean it's an unwarranted bias in the sense we're investigating here.

Evidence from studies like this is limited because they examine hypothetical hiring decisions involving rather generic job applications. Under such circumstances it could be rational to include base-rate information such as gender differences in productivity in one's decisions (whether it is moral or legal to do so is a very different matter). Consider this analogy. Imagine that you run a removal company and your employees have to be fit and strong enough to do a lot of lifting of furniture, boxes, pianos, and so on. You need to hire a new employee. If all you have to go on are job applications that reveal contact details and mundane information about educational qualifications, then choosing only to interview male applicants would be perfectly sensible given that males on average are stronger than females. If, on the other hand, you receive applications from individuals, all of whom describe previous experience working for removal companies and excellent job appraisals and letters of recommendation, then it would be counterproductive to pay any attention to gender. The point is that the more generic and nonspecific an application, the more sensible it is to put weight on statistical signals such as gender. To reiterate, we are talking here about whether decisions are biased or impartial, not about whether they're legally or morally acceptable or fair.

Proving the point, when we look at more realistic hiring decisions, involving much more comprehensive information about the candidate, gender bias against women is much harder to detect. Much of this research has focused on hiring in university departments. In the largest investigation undertaken to date, Wendy Williams and Stephen Ceci asked faculty in a range of science disciplines to evaluate realistic job applicants for a tenure-track position.[12] The application materials were highly realistic and included detailed notes describing the search committee chair's evaluation of each candidate, with comments such as "Z's faculty job talk/interview score was 9.5/10. At dinner with the committee, she impressed everyone as being a confident and professional individual with a great deal to offer the department. During our private meeting, Z was enthusiastic about our department, and there did not appear to be any obstacles if we decided to offer her the job." Of course, the same materials were rated by some faculty under a female identity and by others under a male identity.

Remarkably, female applicants were favored over males by a ratio of about 2:1. This pattern held over both math-intensive fields like engineering as well as less math-intensive ones like biology, the only exception being economics, where no preference for either gender was found. Indeed, evidence on the actual hiring rates of men and women for faculty positions at US universities shows a similar preference for, not against, women (although fewer women apply overall). Between 2002 and 2004, for instance, around 20 percent of all applicants for faculty math positions in universities in the United States were women, but 32 percent of those offered jobs were women.[13] Of course higher education is only one among many employment sectors, and there is a persistent gender pay gap in almost all types of employment. But evidence for the idea that psychological factors such as implicit bias play a role in hiring decisions is distinctly underwhelming.[14] To state the obvious, judgments and preferences can't reflect unconscious biases if they're not even biases.

Similar patterns emerge in other contexts in which gender (or race) is one among many indicators. For example, in scientific publishing, researchers or research teams send their often lengthy and detailed reports to journal editors who solicit evaluations from expert reviewers and decide whether to publish them. These are important decisions as promotion and pay in universities and other research organizations depend hugely on one's track record of scholarly publications. Although women make up less than one-third of the authors of all scientific publications globally, articles submitted to journals by female authors are no less likely to be published than those by male authors.[15] When gender is rendered an irrelevant signal, it is ignored in decision making. It is irrelevant in this case because the decision about whether to publish an article should be (and is) based on its intrinsic quality.

Indeed far from supporting the common (mis)conception that biases rest on inaccurate stereotypes about social groups, the evidence is much more consistent with what has been dubbed the "kernel of truth" hypothesis, which maintains that many of our stereotypes tend to reflect reality, at least to a first approximation. If people generally hold the stereotype that women tend to be more conscientious but less extroverted than men, perhaps this is because it's actually the case—which indeed is what the research findings show. The social psychologist Lee Jussim has argued forcefully that the automatic assumption that stereotypes are inaccurate grossly misrepresents reality.[16]

Unconscious Associations

Alongside this sort of quasi-naturalistic research on implicit bias, there has been much investigation using laboratory tasks, most famously the well-known implicit association test—the IAT. This test for measuring unconscious attitudes, devised by Anthony Greenwald and his colleagues at the University of Washington, takes the form of a simple set of decisions. Imagine that you're shown a series of words one at a time on a computer screen, such as *lovely*, *dirty*, *friendship*, *rotten*, *appealing*, *evil*, *cheerful*, *failure*, and you have to press a left key for positive words (such as *cheerful*) and a right key for negative words (such as *dirty*). The speed with which you make these decisions is recorded. Next you're shown photographs of Black or White people and again required to press one key for Black faces and the other key for White faces. This is all easy.

Now we get to the interesting part. You see a sequence of randomly interspersed words and faces and have to press the left key for negative words and Black faces and the right key for positive words and White faces. It should be intuitive that if you have an unconscious bias against Black people, it will be relatively easy to respond rapidly in this stage of the experiment as the separate decisions (Black—left; White—right; negative—left; positive—right) are compatible in the sense that the left key is used for both negative words and the relatively disliked faces and the right key for both positive words and the relatively liked faces. Certainly your decisions should be faster than in the final part of the experiment, in which you have to press the left key for negative words and White faces and the right key for positive words and Black faces. In this case, the separate decisions are incompatible (the left key is used for negative words and relatively liked faces).

Implicit association tests such as this can be constructed to measure a wide range of biases involving age, sex, disability, attitudes to overweight or obese individuals, and so on, in addition to race. Figure 5.1 illustrates the method when testing for age bias. A substantial majority of individuals are classified as showing apparent bias by these measures. Across over 2.5 million people tested in an online version of the IAT, approximately 70 percent showed a Black-White race bias, and an even greater percentage showed an ageism bias. Black individuals showed no detectable bias, favoring neither race.[17] Advocates of implicit measures like the IAT suggest that they measure very stable, long-term attitudes built up over a lifetime of social interactions embedded in our culture.

Such unconscious biases have an obvious link to the concept of microaggressions, relatively subtle acts constituting insults or slights directed toward members of disadvantaged groups. A taxi driver who drives past a Black person and picks up a White passenger is committing a microaggression, as is someone in the workplace who describes assertive females as shrill but assertive males as strong. Although less overt than direct acts of racism or sexism, the harm these cause is no less real, and many have argued that microaggressions can contribute to adverse mental health. In the past few years, a huge microaggression industry has emerged (it was the 2015 word

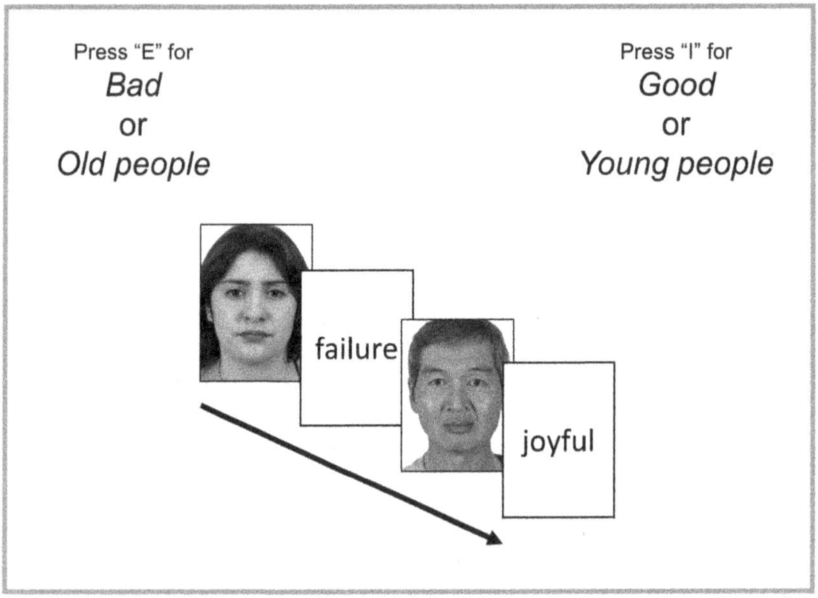

Figure 5.1
Schematic illustration of the implicit association test. This example is for measuring unconscious age bias. The faces and words illustrated here are shown one at a time on successive trials in a randomized sequence. In this consistent phase, participants press one button when a "bad" (negative) word or old face is shown, and a different button when a "good" (positive) word or young face is shown. In another (inconsistent) phase, these assignments are switched so that negative words and young faces are paired with the same button press, and the same for positive words and old faces. The difference in average speed of responding between the inconsistent and consistent phases provides the IAT measure of unconscious bias. (Faces from Debbie S. Ma, Joshua Correll, and Bernd Wittenbrink, "The Chicago Face Database: A Free Stimulus Set of Faces and Norming Data." *Behavior Research Methods* 47 (2015): 1122–1135. https://doi.org/10.3758/s13428-014-0532-5).

of the year, according to one source), but the status of the concept as a subject of scientific investigation is rather dubious.[18] Part of the definition of microaggressions is that they lie in the eye of the beholder: if individuals feel that acts directed at them are microaggressions, then they are. There's nothing wrong with this from the perspective of using the term in our everyday cultural discourse, but a concept that cannot be objectively verified by outside observers is a poor candidate for scientific scrutiny.

Is Unconscious Bias Unconscious?

How do we know that the sorts of biases measured in the IAT or causing microaggressions are automatic and unconscious? Several justifications have been offered. Although people are often reluctant to report their overt attitudes toward disadvantaged groups, there are well-validated questionnaires for doing so. These questionnaires, which include a widely used test known as the Symbolic Racism scale, attempt to avoid such "self-presentation" or "impression management" (in social psychologists' jargon) concerns and solicit genuine attitudes by reassuring respondents that their responses are anonymous. If the IAT were simply another method for measuring explicit attitudes, we would expect to see strong correlations when we compare individuals' IAT scores with their explicit attitudes. This is not what is found, however. Instead, IAT scores tend to show only rather weak associations with measures of explicit attitudes, as would be expected if the IAT is measuring something distinct from conscious prejudice—that is, unconscious prejudice.[19]

There are, however, reasons to regard the IAT as a rather poor instrument for showing that biases can sometimes be unconscious. Indeed, two of the staunchest critics of implicit prejudice, Gregory Mitchell and Philip Tetlock, pulled no punches when they recently wrote:[20]

> It is also difficult to find a psychological construct that is so popular yet so misunderstood and lacking in theoretical and practical payoff. Scholarly discussions of prejudice fail to agree on how implicit prejudice connects to other forms of prejudice; it is unclear whether different measures of implicit prejudice measure the same thing; the meaning of "implicit" in the phrase "implicit prejudice" is contested; and implicit measures of prejudice are no better at predicting behavior, even "microaggression'" (small, barely visible slights), than are traditional explicit measures of prejudice.

What motivates this coruscating criticism? There are many reasons, usually overlooked or even ignored in popular discussions of unconscious bias (in chapter 9, we discuss evidence that this biased evaluation even permeates the discussion of the IAT in introductory psychology textbooks). People seem, for instance, to have sufficient insight into their IAT responses to be able to predict them ahead of time, a capacity that surely implies a considerable degree of conscious access. When shown race IAT items—Black and White faces and positive and negative words—and asked to estimate how easy or difficult they think sorting Black faces with positive words or sorting White faces with positive words would be, people are able to respond with considerable insight.[21] They can also, at least to some degree, deliberately fake their responding in an IAT.[22] Germans, who according to the IAT typically show prejudice against members of the minority Turkish population in Germany, can readily suppress their IAT scores and actually fake pro-Turkish attitudes when motivated to do so. The idea that the IAT is an instrument for measuring those aspects of our biases and prejudices that are beyond our conscious control is seemingly quite incorrect.

This is not the end of the difficulties faced by those who wish to use the IAT as a measure of unconscious bias, however. It appears that the test is—with a striking exception noted shortly—excessively liberal in classifying individuals as biased. Suppose we have some fairly unequivocal indicator of racial bias. Perhaps we film individuals interacting with a Black or White person and score these interactions in terms of friendliness, abruptness, physical proximity, and so on. Individuals who manifestly behave indistinguishably in interactions with a Black or White person show no tendency toward racial discrimination. And suppose we now administer the IAT to such individuals. Naturally, if it's a fair measure of bias (whether unconscious or not) it should reveal no hint of bias. But studies find something very different: that the IAT yields positive bias scores (that is, a preference for White over Black people) even in individuals who show no detectable bias in their real human interactions.[23] It's hard to interpret this as anything other than evidence of bias in a test designed to measure bias.

The exception to this liberal classification of people as biased is that the IAT reliably fails to classify men as being sexist.[24] In IATs designed to measure prejudice by men against women, minimal bias emerges. This is a striking and thus far unexplained anomaly. For defenders of the IAT, it must mean that there is little covert sexism in modern society, a conclusion that few would endorse. Other research shows that in contrast to explicit attitude measures, the IAT is a quite unreliable test in the sense that the scores a given individual achieves when tested twice—across even quite short intervals such as two weeks—tend not to be very consistent.[25]

Although the IAT is by far the most widely used and evaluated test of implicit attitudes, it is far from the only one. Social psychologists have devised numerous other measures with names such as the evaluative priming and affect misattribution tests. If these tests all measure a common underlying propensity—unconscious bias—then they should all yield similar results when large samples of people are tested on all of them. Someone who scores highly on the IAT for racism or ageism should also score highly on the other tests. But the picture is in reality rather different from this ideal. An individual classified as highly biased by the IAT is unlikely to achieve a similarly high bias score on many of the other tests. Overall, correlations between scores on these different measures tend to be quite modest.[26] Without compelling reason to regard any one of the tests as better than the others, this means that in reality, we have no defensible way of accurately measuring unconscious bias. Indeed the pattern of low overlap between the different tests is exactly what one might expect if in reality there's no such construct as unconscious bias and if instead each test is measuring something highly idiosyncratic.

What might such idiosyncrasies of measurement be? In the case of the IAT, there are many possibilities. It might at least in part be measuring the familiarity or salience of the target categories. For a White participant taking a race IAT, Black faces are likely to be less familiar or more salient than White ones. Perhaps it is this difference that causes faster responses for compatible than incompatible trials? It might be generally easier for people to pair familiar and good objects on the one hand and unfamiliar and bad ones on the other, rather than vice versa. This would yield the typical race IAT bias effect but with no reason to attribute it to negative reactions to other-race people. Researchers have obtained clear evidence supporting this possibility.[27]

Significant though all these measurement problems are, an even greater concern is that implicit bias as measured by the IAT seems to have only the most tenuous connection to actual prejudicial or discriminatory behavior.[28] Even if the IAT were a valid and reliable measure of unconscious bias, which it plainly isn't, if this bias appears to have very little to do with overt behavior, then it's hard to argue that it's any sort of cause of how we act toward others. We've already seen that the IAT tends to classify individuals as biased even when they're behaviorally neutral. As Hart Blanton, James Jaccard, and their colleagues have argued at length and in detail, research shows that IAT scores don't do a very good job at all of predicting behavior over and above explicit reports of attitudes.[29]

For example, if one measures the racial attitudes of a large number of individuals using standard and well-validated questionnaires such as the Symbolic Racism scale, these predict discriminatory behavior reasonably well. If implicit bias is distinct from explicit bias, then it should follow that including scores from the IAT ought to improve even further the ability to predict discriminatory behavior, because deep-rooted unconscious biases that will fail to be detected on a self-report questionnaire will nonetheless leak through into actual behavior. But there is no research that convincingly demonstrates this. On the contrary, the added value of IAT scores is close to zero in many analyses.

The same general conclusion follows from research on the effects of interventions designed to change implicit bias. People can be given strong arguments explaining why racial bias is unacceptable, and these arguments can have some effect on reducing automatic bias, as measured by the IAT. Worryingly for defenders of the IAT, however, this research suggests that these interventions have a negligible effect on actual prejudicial behavior. Patrick Forscher and his colleagues, who conducted an extensive meta-analysis of research on this issue, noted that "there is little evidence in our data that is consistent with a causal relationship between automatically retrieved associations and behavior."[30]

The point is not to dismiss all of this research out of hand, and certainly not to dispute the real harm that discrimination causes to many groups. Rather, it is to emphasize that almost all of the scientific evidence for unconscious bias is controversial and open to alternative interpretations. Only when the facts have been established more convincingly one way or the other will we be able to draw firm conclusions about the specific significance of unconscious bias, in contrast to more general problems of discrimination, in the workplace and other settings.

Nor does it help when workplace training programs to counter unconscious bias lack scientific rigor. At one prestigious university (which shall remain nameless), the mandatory unconscious bias training video claims that White interviewers sit farther away from and end interviews sooner with Black candidates than White candidates, and it cites two studies documenting this claim.[31] And the world's oldest academic body, the Royal Society, states in its unconscious-bias video that people pay more attention to male than female voices.[32]

How convincing is this evidence? Both of the studies about interviews were conducted over forty years ago and used student participants in simulated interview settings, so their relevance to modern real job interviews is negligible. Moreover, in the study that reported that the distance between an interviewer and interviewee was greater when they were from the same race than when one was White and the other Black, it was the interviewee (not the interviewer) who chose the seating distance, so the effect cannot be evidence of interviewer bias, and the interview was about political attitudes rather than a (mock) job interview. Likewise, there is little support for the claim that people pay more attention to male than female voices. What the cited research actually found was something entirely different: that people prefer to vote for both men and women with deeper voices.[33] There is a strong impression of confirmation bias in these examples: the creators of these training videos are so convinced that unconscious bias exists that they give unjustified weight to any evidence, however dubious, that can be claimed to demonstrate it.

The sense in which unconscious bias has been assessed in this chapter—irrational preferences or behaviors—can be contrasted with a different and less controversial use of the term. This relates to ways in which we can be unaware or oblivious of the effects of our behaviors on others. To give an example, gendered language (using terms like *businessman* and *chairman*) is ubiquitous. Simply by habitually following social conventions or norms in language and culture, each of us may unwittingly reinforce disparities across gender, race, and so on. When an advertising agency uses photographs of thin models, it is contributing to our culture's idealization of a certain body shape. Many people would agree that in an equal society, such biases should be pointed out and discouraged, but there is no necessary link between using the word *chairman* and having unconscious sexist attitudes. One can follow prevalent conventions without being psychologically biased.

The societal implications of the current climate around unconscious prejudice can hardly be exaggerated. Consider the following conclusion from a recent review:

> A growing body of research suggests that similar to the general US population, most HCPs [health care providers, that is, doctors] across multiple levels of training and disciplines have implicit biases against Black, Hispanic, American-Indian and dark-skinned individuals (Maina et al., 2017).

At face value, this is a gross indictment of almost the entire medical profession, stated with virtually no acknowledgment of the many reasons to be cautious, if not downright skeptical, about the validity of tests like the IAT to accurately probe our unconscious attitudes. We can all agree that highlighting situations in which bias may occur is a worthy cause, that racism, sexism, ageism, and so on are immoral and deplorable, and that society and institutions have a duty to put policies in place to prevent discrimination and intolerance. But the evidence available to date doesn't even come close to proving that most of us walk around with unacknowledged and unconscious biases in our heads.

6 Think, Blink, or Sleep on It?

In order to "get over" the "uncertainty that perplexes us" when we are faced with important, complex decisions, the American politician and polymath Benjamin Franklin advised his friend the British scientist Joseph Priestley thus:[1]

> My way is to divide half a sheet of paper by a line into two columns; writing over the one Pro, and over the other Con. Then, during the three or four days consideration, I put down under the different heads short hints of the different motives, that at different times occur to me, for or against the measure . . . I find at length where the balance lies; and if, after a day or two of further consideration, nothing new that is of importance occurs on either side, I come to a determination accordingly. . . . And, though the weight of reasons cannot be taken with the precision of algebraic quantities, yet when each is thus considered, separately and comparatively, and the whole lies before me, I think I can judge better, and am less liable to make a rash step.

It was July 1772, and Priestley was grappling with whether to accept the offer of becoming Lord Shelburne's librarian. He reasoned that Shelburne would be an influential patron and that the salary was good. And yet the upheaval involved in leaving his productive and comfortable life in Leeds made it a tough decision. In the end, after almost six months of deliberation, Priestley accepted the offer.

Half a century later, one of the greatest intellects of the modern era was faced with an even more momentous decision. Whether Darwin knew of Franklin's technique history does not relate, but as figure 6.1 illustrates, he adopted a similar method. The figure shows a page from Darwin's journal with two columns headed "Marry" and "Not Marry" separated by the statement, "This is the question." Below each are various pros, such as companionship in old age, and cons, such as disruption to his work, which entered

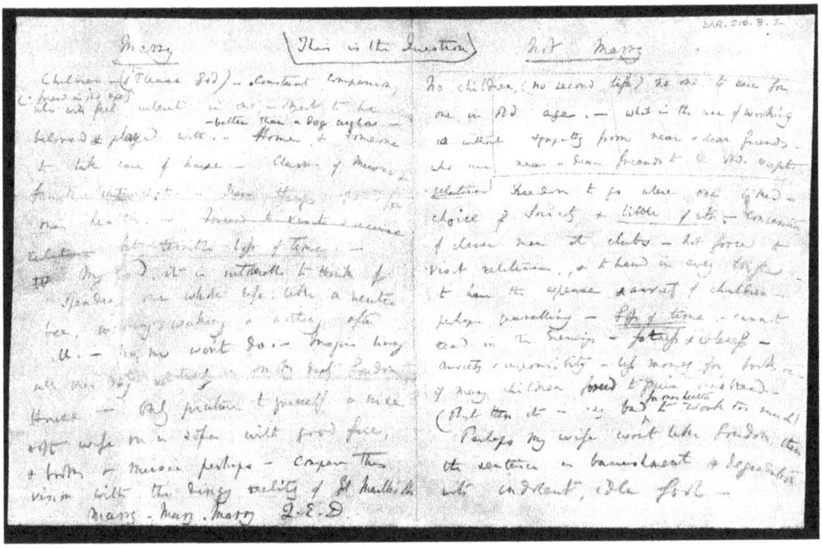

Figure 6.1
A hand-written note by Charles Darwin showing how he grappled with his decision about marriage. At the top of the left page the word "Marry" is underlined, with "Not Marry" underlined on the top of the right page and "This is the question" in boxed text in the center. Reproduced by kind permission of the Syndics of Cambridge University Library. (Image reference: DAR 210.8: 2.)

into Darwin's cost-benefit analysis of getting hitched. His conclusion (visible at the bottom of the left page): Marry, marry, marry! QED.[2] Charles and his cousin Emma Wedgwood tied the knot on January 29, 1839, had ten children, and were married for over forty years. So at least in terms of family and longevity, getting married appeared to have been a good decision.

Franklin's advice to Priestley and Darwin's strategy seem intuitively sensible. When faced with a difficult decision, we should carefully consider the evidence before us, weigh things up, and try to settle on the best course of action. Indeed, the essence of what Franklin called his "moral" or "prudential algebra" is found in many of the formal descriptive and prescriptive approaches to decision making.[3] This advice to rely on explicit, conscious thinking also resonates with our central argument: there is no free lunch when it comes to tricky decisions; you have to do the thinking. The alternative, delegating decisions to the lower reaches of the iceberg and hoping that the unconscious will decide for us, is, we argue, misguided. As we have seen in the preceding chapters, evidence for the ghost in the machine helping us to decide when to decide is scant; the idea that ripples of activation in our subconscious mind have impacts on our behavior does not bear scrutiny, and evidence for truly unconscious biases seems difficult to find. Moreover, the study of information leakage shows how acutely sensitive we are to the ways in which we are asked to do things.

And yet a persistent idea in popular conceptions of how the brain and mind work is the notion that thinking can occur outside awareness and that, indeed, harnessing this power of the unconscious brain can lead to better outcomes than striving to think. The argument comes in two guises: that we should go with our gut reaction (blink, or not think) or that we should delegate cognitive activity to an unconscious part of our brain and take a break (sleep on it). In this chapter we explore the evidence for blinking, thinking, and sleeping on it and ask is there a "best" way to decide?

Can Too Much Thinking Be a Bad Thing?

> Well of course too much is bad for you, that's what "too much" means you blithering twat. If you had too much water it would be bad for you, wouldn't it? "Too much" precisely means that quantity which is excessive, that's what it means. Could you ever say "too much water is good for you"? I mean if it's too much it's too much. Too much of anything is too much. Obviously. Jesus.
>
> —Stephen Fry and Hugh Laurie, *Doctor Tobacco*[4]

A common argument for "going with your gut" or "sleeping on it" is the claim that if we keep deliberating on a decision, we might end up in a kind of analysis paralysis. The idea is that our conscious brain has capacity limitations: it can only hold in mind the magic number of 7 (plus or minus 2) pieces of information at a time, and thus is hopelessly hobbled when it comes to complex decisions.[5]

A famous experimental example of this too-much-thinking effect involves strawberry jam.[6] The setup was as follows. Participants were brought into a laboratory and were asked to taste five different jams lined up on a table in front of them. After tasting the jams, they were asked to rate how much they liked each one. However, before making the ratings, half of the participants were asked to write down their reasons for liking or disliking each of the jams, while the other half (a control group) listed reasons for a completely unrelated decision: choosing their university major. The experimenters had carefully selected the jams to be representative of a spectrum of quality according to trained experts. The crucial question was whether the participants who were asked to think and provide reasons for their preferences or those who judged without justification ended up with ratings more similar to those of the experts.

The results were intriguing. To obtain a measure of similarity, the experimenters looked at the correlation between the ranks of the students' ratings and the experts' ratings. In the control condition, this correlation was a pretty high 0.55 (where 0 is no relationship and 1.0 is a perfect relationship; in psychology, any correlation above about 0.3 is considered informative). But contrary to received wisdom, the correlation between experts and students who had listed their reasons for liking or disliking the jams was only 0.11—in other words, pretty much no relationship. What happened?

One possibility is that the process of introspecting about why we like a particular jam makes it harder for us to figure out the real reasons because they lie outside our conscious awareness. The suggestion is that although we typically are able to weight relevant information appropriately, for some decisions, it leads to a focus on a subset of reasons that are accessible or plausible (for example, the chunkiness or tartness of the jam) that may not have directly influenced our initial reaction. As a result, this subset of reasons may receive greater (inappropriate) weight than other possible unarticulated reasons, diminishing the quality of the final preference or choice.

One might argue that this suboptimality of introspection only happens when the decision is one that does not lend itself to verbalization. Is it always possible to put into words why something tastes better or worse than something else? Clearly the expert tasters may be able to because they have developed the vocabulary that allows for subtle distinctions. We all marvel at wine connoisseurs for the decorative and inventive language they employ, and expert jam tasters are no different. But even with this factor in mind, there are other features of the jam experiment that should lead us to pause before concluding that thinking leads to suboptimality.

First, it turned out that there was a very high correlation between the liking for each jam expressed in participants' reasons and their explicit liking rating. So, for example, if a participant reasoned that they really liked the fruitiness of a particular jam, then this was reflected in their subsequent rating. Thus, the participants were internally consistent and could access the proximal bases for their choices even if the reasoning process led them to divert from the "optimal" reasons provided by the experts. This internal consistency may well have been driven by a demand characteristic of the experiment: if participants thought that they might be evaluated for consistency between their reasons and their ratings, then they might have strived to line these up.

A second feature of the experiment was that most of the participants had to taste all five jams first and then provide their reasons, retrospectively, for each one. This delay between tasting and justifying introduces the potential for sensory as well as memory-based interference, which may have pushed people toward relying more on those accessible attributes. Intriguingly, a footnote in the original study lends some weight to this possibility: a few participants in an initial version of the experiment provided their reasons consecutively after tasting each jam, and they showed a stronger correlation with the experts' rankings. Unfortunately, there were too few participants in this version (only five), and so we cannot draw any strong conclusions from this finding, but the results are certainly suggestive.

The reason for dwelling on the details of this experiment is that the conclusions have gained folkloric status. The jam-tasting study has become synonymous with the idea that "too much thinking is bad," which in turn leads to the idea that relying on the unconscious mind is good, without always properly considering the evidence. Indeed, the authors, Timothy Wilson and Jonathan Schooler, were more conservative in their original conclusions, stating (in a variation of the famous Socratic advice) that "at least at times, the unexamined choice is worth making."[7]

This conclusion is fine as far as it goes, but it leaves us hanging in two important ways: we still don't know which choices require additional reasoned thought, and, even more frustrating, it tells us nothing about where our initial reactions come from. Recall that the ratings provided by participants in the control group lined up pretty well with those of the experts. This overlap might suggest that the experts and the control group participants engaged in similar processes when making their judgments. But what were these processes? If not thinking, evaluation, and attribution, then what? One cannot simply remove the explanation by stating that our reactions bubble up from some inaccessible state. Or perhaps, as we will discover next, you can?

Can *NOT* Thinking Be Better Than Thinking?

History is replete with examples of dramatic insights or great works of music and literature simply popping into people's minds. The German chemist August Kekulé had a vision of a snake biting its tail, which supposedly revealed the ring structure of benzene. The French mathematician Henri Poincaré restricted his working hours to allow his unconscious mind to work on tricky math problems in the downtime. Samuel Taylor Coleridge, the celebrated English poet, apparently wrote his best-known poem, "Kubla Khan," while asleep.[8] The melody for "Yesterday," perhaps the Beatles' most famous song, came to Paul McCartney in a dream. Sir Paul later claimed that he woke up, fell out of bed (perhaps dragged a comb across his head . . .), stumbled to his nearby piano, "found out what key I had dreamed it in . . . and I played it."[9]

Such accounts, and there are many more, seem miraculous and appear to fly in the face of our claim that there are no murky depths from which fully formed thoughts, ideas, hypotheses, or solutions arise. How can we reconcile these famous examples—which all probably resonate with similar, though probably less momentous, experiences of our own (the crossword clue coming to us in the middle of the night; finally remembering the name of the person we saw on the bus or where we'd put our keys)—with the kind of data we would need to go beyond what are, after all, simply memorable anecdotes? And as the saying goes, the plural of anecdotes is not anec*data*.

This point is worth a little further consideration. In essence, it is an instance of something we will discuss in more depth in the second part of the book—the so-called *file drawer problem* (see chapter 9). We hear about the lottery winner who dreamed about the winning numbers, but we never hear from people who dreamed about losing ones. This seems so obvious that it is hardly worth pointing out. What news organization would run a story about the person who dreamed about the wrong lottery numbers? But if we focus only on the surprising, memorable, positive instances, it is difficult to evaluate the claims for miraculous influences. How often did Sir Paul roll out of bed, hammer out some chords, and then think, "Nah, nothing happening there"? We'll never know. More generally, given the sheer number of times we dream and the myriad contents of those dreams, chance alone will lead some of them to appear consequential. But such statistical flukes should not be the basis for our theories of how the mind works.

The limitations of anecdotal reports—however intuitively plausible they might seem—requires us to reenter the psychology laboratory and look for direct evidence that not thinking about a problem or a decision can actually be beneficial. Dutch social psychologist Ap Dijksterhuis and his colleagues have done exactly that in pursuit of their unconscious thought theory. They claim that being distracted from a decision allows unconscious thought processes to help us achieve a better outcome. The benefits of this process are argued to be strongest when a decision problem is complex—those with multiple options and attributes—because unconscious thought does not suffer from the capacity limitations that hobble conscious thought.

So what are these complex decisions that benefit from a period of unconscious thought? In the lab, researchers have used hypothetical versions of decisions like choosing a new car, an apartment, or a roommate. Take the example of deciding what car to buy. In the standard experimental paradigm, participants are presented with information about three or four fictional cars (for example, a Hatsdun, a Kaiwa) described by ten or more attributes (mileage, trunk space) and are asked to choose the "best" one. Each car has a different number of good and bad attributes (perhaps the Hatsdun might have good mileage, trunk space, lots of cupholders but a high price). When this information is presented to participants, things get a little odd. Instead of providing the list of attributes in a convenient table, allowing for an easy comparison of options, the information appears sequentially, one at a time in the middle of a computer screen and typically in a random order about the three or four different cars.

The next part is where the experimenters try to mirror the think, blink, or sleep-on-it methods. One group of participants is told that they will be asked about their car choice later, but first they need to solve some anagrams. This period of distraction is claimed to facilitate unconscious thought. In essence, you allow your brain to sleep on the car choice decision for a few minutes (unconscious thought takes care of it for you) while your conscious brain grapples with working out what LOPTI might mean. Those asked to "think" are also told that they will need to choose a car a little later but that they should use the next few minutes to really consider their choice carefully. They have to do this, however, by relying on their memory for the randomly presented car attributes; they are not allowed to review the information. Finally, the "blinkers" are asked for an immediate car choice as soon as the final attribute has disappeared from the screen.

Who makes the best decision? The sleepers, the thinkers, or the blinkers? The answer might come as a surprise: distraction appears to lead to better choices than either conscious thought or an immediate decision. For example, in one early study published in the highly prestigious journal *Science*, 60 percent of participants chose the best car after being distracted compared to only 25 percent following conscious deliberation.[10] Proponents of unconscious thought theory explain this surprising pattern of results along the following lines. The bombardment of all forty pieces of information about the cars (four cars each with ten attributes) in a random order is simply too much for us to comprehend in a methodical, conscious manner, and so when we are asked for an immediate decision or, worse, forced to cogitate on a loose collection of impressions (Was it the Hatsdun or the Kaiwa that had lots of cupholders?) we fail miserably. Note that a 25 percent choice rate of the best car when there are only four options means that people chose randomly. Unconscious thought is claimed to have increased capacity and superior information weighting relative to conscious thought. Thus, while our attention is held by anagrams, the unconscious acts in the background to organize, weight, and integrate information in an optimal fashion, ready for the answer to bubble up once we are asked.

This kind of explanation is anathema to our argument: nothing can bubble up from the depths of the iceberg because there is no active cognition—of the kind that leads to decisions, judgments, and preferences—occurring below the limen of awareness. So how do we reconcile this apparent advantage for unconscious thought with our central proposal? One route to reconciliation is to explore alternative explanations of the effect that do not require recourse to ill-defined unconscious processes. For example, how do we know that unconscious thought leads specifically to superior weighting of information? In many of the experiments, the importance of attributes, and thus what constitutes the "best" car, is predefined by the experimenter. Often this is done in an implausible manner by, for example, deeming the number of cupholders in a car as important as the fuel economy. With these experimenter-defined weighting schemes, it is impossible to know if the "best" choice is indeed the one favored by all participants. (I might think that cupholders are twice as important as fuel economy; you might think the opposite.)

In fact, subsequent studies that asked participants for importance ratings for the attributes (for example, "How important are cupholders for making your decision?") found that regardless of the mode of thought—conscious, unconscious, or immediate—the majority of participants chose the option predicted by summing up their subjective importance ratings. In other words, I choose what I like (not what the experimenter thinks is best), and it does not matter how you make me think about it.[11]

A second route to reconciling apparent unconscious thought advantages with our perspective is to question the reliability of the whole enterprise. Following the initial high-profile publication of the cars study in *Science*, a large number of researchers engaged in further investigations to see just how far the deliberation-without-attention idea could go. It is fair to say that the results of these studies were mixed: many researchers (us included) found that they could not replicate the basic effect—often finding no differences in terms of the choices made by the thinkers, blinkers, and sleepers.[12]

The final nail in the coffin for the theory was a large experiment with almost four hundred participants that attempted to determine once and for all the circumstances under which *not* thinking might be better than thinking.[13] The team, also Dutch (but not involving Dijksterhuis and his colleagues), included a variety of conditions to try to pinpoint the sweet spot for unconscious thought. For example, some previous work had suggested that the distraction task engaged in during the period of not thinking needs to be "just right" in terms of its difficulty—not too hard but not too easy, the idea being that unconscious thought might need a little bit of attention, thereby making lighter distraction tasks more fruitful than taxing ones. Thus, among many other things, this large experiment included groups given either difficult anagrams or an easy word search problem during the distraction period. Did this make any difference to the quality of the decisions? None whatsoever. The bottom line from this study was that regardless of the mode of thought, the type of distraction task, the time for (non)deliberation, the complexity of the decision, the participants' goal—the list goes on—60 percent of participants on average chose the "best" option. There was no advantage of unconscious thought. Moreover, a meta-analysis (a method for combining results across experiments that we return to in part II) indicated that existing evidence for unconscious thought came from studies with relatively small numbers of participants, thus casting doubt on the reliability of the effect.

Where do these investigations of jam tasters and hypothetical car buyers leave us? Are we any closer to knowing how best to decide? Before answering this question, let's turn to another source of evidence that might shed some light. The studies we've discussed so far in this chapter have tended to focus on what people do after they've been given some information (jams to taste, cars to choose) and have asked how much time they should devote to (not) thinking about the options in front of them. But what about a more fundamental question: How much information do we need in the first place for a good decision? Should we follow Franklin's advice and keep searching until "nothing new that is of importance occurs on either side" of the ledger, or should we rely on a brief glimpse?

Blink, Don't Think?

Recall the study about the cooperativeness of rural Senegalese men and women in chapter 5. Researchers found that a brief look at the face of a male from a different culture is sufficient to extract valid information about that person's cooperativeness. A glimpsed photograph facilitated a seemingly complex inference about pro-sociality. This surprising finding seems to reveal something fundamental about our ability to make accurate judgments on the basis of very thin slices of information. So perhaps the powers of the unconscious are revealed not when our thinking per se is curtailed but when the evidence we have before us is scant and impoverished. Only then do the processes at the bottom of the iceberg come into their own.

"Thin slicing," proposed by social psychologist Nalini Ambady (and popularized by Malcolm Gladwell in his best-seller *Blink*), captures the idea that very brief observations of behavior can give rise to surprisingly accurate inferences.[14] The typical study involves participants watching or listening to brief clips of one or more people about whom they are then asked to make a judgment.

For example, one early study showed participants ten-second clips of thirteen college teachers and asked them to rate each teacher on a variety of characteristics (likable, dominant, confident, warm).[15] These ratings were then compared to an assessment of the overall effectiveness of the teacher provided by students who had been taught by the same teacher for an entire semester. Remarkably, these ten-second clips were enough for the observers to get a good sense of the teachers. The ratings for traits like optimism, enthusiasm, and confidence correlated very strongly (more than .70) with the end-of-semester judgments. A subsequent study showed that reducing the slice down to two seconds did not make observers any worse.

For Gladwell, these kinds of demonstrations are evidence of the "ability of our unconscious to find patterns in situations and behaviors."[16] But is it really the unconscious at work here? Let's dig a little deeper into the details.

In his book, Gladwell discusses (at some length) the work of John Gottman and colleagues on judging the success or otherwise of marriages on the basis of brief videos of couples interacting. A typical video might feature a fifteen-minute discussion about a couple's new dog; that is, couples are not directed to talk about their marriage, but the topic they choose might reveal some insights into how things are going. Careful inspection of these videos reveals hints that can apparently predict marriage longevity with high accuracy. But here's the thing: it is careful inspection, not a mere glimpse or blink. In a telling passage, Gladwell notes that when he tried to judge the success or failure of couples in a set of videos sent to him by Gottman, he did no better than flipping a coin. If the unconscious is so talented at sifting the wheat from the chaff, of doing the organizing and weighting of information, then why was Gladwell such a poor judge?

The answer in part lies in the fact that to become a marriage-longevity whisperer requires knowing what to look for in the videos. Gottman uses his specific affect coding system (SPAFF) to document all of the emotional tics and traits that can be gleaned from the verbal and non-verbal behavior of the couple. Far from an untrained eye viewing a "thin slice," this coding system is a painstaking attempt to work out the ratio of positive and negative emotions displayed by the couples. Only with this information carefully recorded can one attempt to predict longevity.[17]

What about naive observers like Gladwell? Can they do better without full-blown training on the SPAFF? One study suggests they can, but again, not just from a glimpse.[18] The study (which had only five raters) gave each participant an emotion checklist (not as in depth as the SPAFF but still a pointer as to what to look out for) and then conveniently divided the ten-minute interactions into thirty-second segments. Raters watched each segment twice—once to focus on the man's behavior and once on the woman's behavior. These pooled ratings were about 80 percent accurate in predicting marriage success. So, yes, with a list of what to look out for and the time to reflect and record the behavior of each member of the couple, raters do appear to pick up some relevant features for predicting marriage longevity. But this hardly sounds like the unconscious finding patterns in situations and behavior.[19]

Nalini Ambady's study of the college teachers had a similar setup. Participants were told what to look for—whether the teacher is confident, dominant, attentive, and so on—and were given time to record their ratings of each clip before the next one was shown. They also had three attempts at it:

three ten-second clips for each teacher. This is not to belittle the results; it is simply to question what role the unconscious plays in their explanation.

In a follow-up to the teacher rating study, reported many years later, Ambady tried to find some more direct evidence for the mechanism underlying the success of thin slicing.[20] She reasoned that if thin slicing reflects nonconscious, automatic processing, then two predictions follow. First, increasing cognitive effort or load while people watched the video clips should not affect performance. Second, encouraging deliberation such as providing reasons for one's ratings should make people less accurate. The logic here is similar to that of the jams and car choice studies. If thin slicing taps processes outside our awareness, then occupying conscious thought with additional cognitive load won't make judgments worse; according to Ap Dijksterhuis, they might even improve. And introspecting on why a teacher is effective might lead you to focus on irrelevant details that cloud your judgment.

To test these ideas Ambady compared four separate groups of participants. One group just made a rating of effectiveness; a second had to count backward in 9s from 1,000 while watching the clips; the third group spent one minute listing the reasons for their ratings following each clip; and the final group sat for one minute between seeing the clip and making their rating (this was a control group to assess whether a simple delay—rather than generating reasons—would have an effect). What happened? The raters who had to report their reasons did much worse than the other three groups, but the immediate, distracted, and delayed groups were all about as accurate as each other. There is certainly no evidence from this experiment that being distracted makes you better—so no support for unconscious thought proponents—but in line with the jams study, it seems that introspection can sometimes hurt. Is this then evidence for an unconscious influence? In reflecting on these results, Nalini Ambady concluded, "The present work does suggest that sometimes it is dangerous to think too much—at least while evaluating others in a familiar domain."[21] This conclusion echoes Wilson and Schooler's claim that at times, "the unexamined choice is worth making." The question is, Why?

Is Intuition Unconscious?

The preferred account seems to be that these too-much-thinking effects are consistent with people lacking conscious introspective access into the "true" bases for their attitudes and subsequent choices. But nothing in these experiments necessitates that conclusion. As we've seen, the key feature of these studies is that participants who are invited to give elaborate reasons end up with choices that are less aligned with those of someone making an impressionistic choice. While such studies support the idea that preferences are constructed, labile, and influenced by deliberation, they surely do not force the conclusion that some influences on choice lie outside awareness. Moreover, there are precious few studies demonstrating that giving reasons causes people to make objectively worse rather than simply less aligned choices.[22]

Choices made intuitively and ones accompanied by an analysis of reasons are, we contend, accompanied by awareness of the proximal basis for that choice. The fact that this proximal basis might not be the same in the two cases does not imply that the unexamined choice was mediated by an unconscious process. We heard about this idea of a proximal basis for choice in chapter 3 when discussing the diners on a blind date. Your choice of a garlic dish from a menu is driven by the proximal belief that it is healthy. There may also be some distal influence that causes this proximal belief—your mother told you it's a good source of antioxidants when you were a child—but that has no bearing on your decision in the moment.

But hold on. Are we having our cake and eating it too? We've said that there is no evidence here for influences from the depths of the iceberg, but we are still describing some choices as being made intuitively while others are the product of deliberation. Isn't intuition simply an unconscious process in a different guise?

No. Herbert Simon, the profoundly influential Nobel Prize–winning economist, computer scientist, and psychologist, famously wrote that intuition is "nothing more and nothing less than recognition."[23] In a similar vein, Albert Einstein once noted that "intuition is nothing but the outcome of earlier intellectual experience."[24] In other words, intuition is something we can rely on when we are in a highly familiar domain where we recognize what to do. Some decisions may appear subjectively fast and effortless because they are made on the basis of recognition: the situation provides a cue (for example, no clouds in the sky), the cue gives us access to information stored in memory (rain is unlikely), and the information provides an answer (don't take an umbrella). When such cues are not so readily apparent or information in memory is either absent or more difficult to access, our decisions shift to become more deliberative. The two extremes are associated with different experiences. Whereas deliberative thought yields awareness of intermediate steps in an inferential chain and of effortful combination of information, intuitive thought lacks awareness of intermediate cognitive steps (because there aren't any) and does not feel effortful (because the cues trigger the decision). Intuition is, however, characterized by feelings of familiarity and fluency. Again, the simple point is that in neither situation do we need to posit magical unconscious processes producing answers and solutions from thin air (or murky depths).

But what about Paul McCartney and his dream about "Yesterday"? Surely that's solid (anecdotal) evidence that fully formed solutions can emerge from the unconscious? Well, not really. It turns out that McCartney did not record "Yesterday" for another eighteen months after his dream; it took that long to work it up into the classic that it became. It started off being known as "Scrambled Eggs"—with the oh-so-catchy lyric: "Scrambled eggs. Oh my baby how I love your legs." What came to Paul in his dream (and perhaps to many of those other famous beneficiaries of so-called insight) was not so much the fully fledged song as the suspicion of a good melody.[25]

We started this chapter with a question: Is there a best way to decide: blink, think, or sleep on it? Sleeping is good for lots of things, but making tricky decisions does not seem to be one of them, so we can scratch that one (and, incidentally, the same can probably be said for general anesthesia, a much more potent method for switching consciousness off; evidence that the anesthetized brain is capable of meaningful cognitive activities is decidedly slender).[26] Blinking, or going with our gut, should be reserved for domains in which we have a high degree of familiarity. Judging emotions and personalities from faces and brief personal interactions may be one of those domains because we are such social animals. But even then, it seems we need to know what to look out for and have time to record and reflect on information. Thinking, it turns out, can be pretty useful too, but we have to bear in mind that focusing closely on the reasons for our choices can cause those choices to change, as in the jams example. Thinking can appear subjectively fast and intuitive or slow and deliberative, but this distinction has no bearing on the involvement of factors outside our awareness. In fact, as we will discover in the next chapter, although the notion of dual systems of thinking has become pervasive, it does much more harm than good.

7 One, Two, or More? System(s) of Thinking

> People make most judgments and most choices automatically, not deliberatively: we call this "thinking automatically."
>
> Over the past few decades, evidence has mounted that automatic thinking cuts across wide swathes of human behavior to the point that it can no longer be ignored. The anomalies . . . are not minor and scattered. They are systematic regularities that can be of first-order importance for health, child development, productivity, resource allocation, and the process of policy design itself.
>
> —*World Bank Report: Mind, Society and Behavior* (2015)

Consider the following question:

If a bat and a ball cost $1.10 in total and the bat costs $1 more than the ball, how much does the ball cost?

If you are like many people, your immediate answer would be "10 cents." You'd be wrong. Think a little more, and you'll see why.[1] This fluent but incorrect answer is supposedly the product of the kind of automatic thinking described in the opening quote from the World Bank's *Report into Mind, Society and Behavior*. The fact that the World Bank is concerned about this kind of thinking illustrates just how prominent and influential the notion of anomalous judgments arising from flawed, fast thinking has become. Similar statements can be found in reports from institutes of medicine, government departments, and countless consultancies spruiking their tools for overcoming the biases that arise from an overdependence on automatic thinking.[2]

At its core, dual-systems thinking appeals to the idea that mental processes can be dichotomized and compartmentalized into different boxes that house styles of thinking that in some way align or coalesce. Table 7.1 provides an illustrative example of the way in which mental processes get

Table 7.1
An example of dual systems of thinking.

System 1	System 2
Pragmatic	Logical
Fast	Slow
High capacity	Capacity limited
Nonconscious	Conscious
Automatic habits	Controlled
Associative	Rule based
Emotional (hot)	Unemotional (cold)
Independent of cognitive ability	Correlated with cognitive ability

Note: Attributes of systems are clustered according to the properties of apparently distinct and discrete modes of thinking. The evidence in this chapter suggests such dichotomies are simplistic at best and potentially harmful for understanding how the mind operates.

lumped together in these frameworks.[3] Broadly, system 1 captures the kind of automatic thinking that the World Bank is concerned about. System 2 is the slower, deliberative system, often caricatured as *homo economicus*—the rational, omniscient individual.

Debate about dual systems often appeals to intuitions and philosophical speculation, but these discussions typically avoid detailed reflection on concrete evidence. There is in fact a wealth of evidence that researchers have collected and interpreted over recent years as providing firm support for the existence of systems 1 and 2. We will certainly not attempt to review all this evidence here but will go into two examples in some detail. This is important: we should never build our theories about the mind purely on intuition, because on close examination, many intuitions can be revealed as mistaken. One of these examples concerns automaticity and the idea that some mental processes, such as accessing a word's meaning from its printed form, take place without the need for conscious voluntary control and cannot be prevented. But first we will look at evidence that logical thinking can be separated from intuitive thinking. The following example necessarily requires a rather extensive explanation, but the case allows some absolutely fundamental issues and assumptions about whether the mind is composed of two (or more) distinct systems to be investigated.

Belief Bias?

The case is based on a highly influential study conducted by a prominent advocate of the system 1/system 2 distinction and expert in human reasoning, Jonathan Evans, and involves a famous example of apparent irrationality in thinking, *belief bias*.[4] This refers to the common finding that when we judge the soundness of an argument, our judgments tend to be biased by how well the conclusion matches our prior beliefs and expectations. We all believe that ostriches can't fly so if given the argument:

All birds can fly.

Ostriches are birds.

Therefore ostriches can fly.

we judge the argument to be unsound. In fact the argument is perfectly valid; it's the premise that "all birds can fly" that's at fault. Purely from the perspective of judging whether the conclusion is logically valid given the premises, the believability of the conclusion and whether it fits with our knowledge of the world is irrelevant. And clearly people can override this bias. We can all see, given suitable time for reflection, that the argument is perfectly sound. Belief bias is closely related to *confirmation bias*, the tendency to actively search out for information that matches our prior beliefs and expectations.

But what happens when we are unable to reflect adequately on the validity of an argument because our conscious, reasoning mind is unavailable or occupied with something else? This is the question Evans asked. Participants were required to judge the validity of syllogisms such as this, rather harder than the one above:

No healthy people are unhappy.

Some astronauts are unhappy.

Therefore some astronauts are not healthy people.

It was made clear to participants that their task was to judge whether the final conclusion followed logically from the two preceding premises. It is a task about logical validity, not about whether the conclusion happens to be true or false in the real world. The study of syllogisms like this goes back to Aristotle, and they can (as this one illustrates) be very difficult. In fact, it is valid. According to the first premise, the set of healthy people includes none who are also members of the set of unhappy people. The second premise says that some astronauts are in the set of unhappy people, so these can't be at the same time in the set of healthy people. Hence, at least some astronauts are outside the set of healthy people. It is worth reiterating again that logical validity has nothing to do with the actual truth or falsity of any of the statements. Clearly there are not many unhealthy astronauts in the world.

Although actual truth or falsity is irrelevant to validity, it is a factor that Evans manipulated in the experiment, as can be seen from the examples in table 7.2. Two of the syllogisms have quite believable conclusions, while the other two have relatively unbelievable ones. At the same time, the logical validity of the conclusions was varied, two being valid and two being invalid. For example, the syllogism in the bottom left of the table (*no healthy people are unhappy; some astronauts are unhappy; therefore some healthy people are not astronauts*) is invalid because there is nothing in the premises that rules out the possibility that all healthy people are astronauts, in which case the conclusion does not hold.

The other crucial factor that Evans varied was time pressure: for one group of participants, there was no time pressure to make their decisions, while a second group had to make their validity judgments within five seconds. The key findings are shown in figure 7.1. For participants who completed the task under no time pressure, both validity and believability affected their likelihood of judging a syllogism as valid. This is shown in the solid black bars, which indicate that the validity of the arguments made a substantial difference to the likelihood of their being endorsed (the second and fourth bars from the left—the valid ones—are higher than the first

Table 7.2
Examples of the syllogisms used by Evans and Curtis-Holmes (2005)

	Believability of conclusion	
Validity	Believable	Unbelievable
Valid	No astronauts are unhappy. Some healthy people are unhappy. Therefore some healthy people are not astronauts.	No healthy people are unhappy. Some astronauts are unhappy. Therefore some astronauts are not healthy people.
Invalid	No healthy people are unhappy. Some astronauts are unhappy. Therefore some healthy people are not astronauts.	No astronauts are unhappy. Some healthy people are unhappy. Therefore some astronauts are not healthy people.

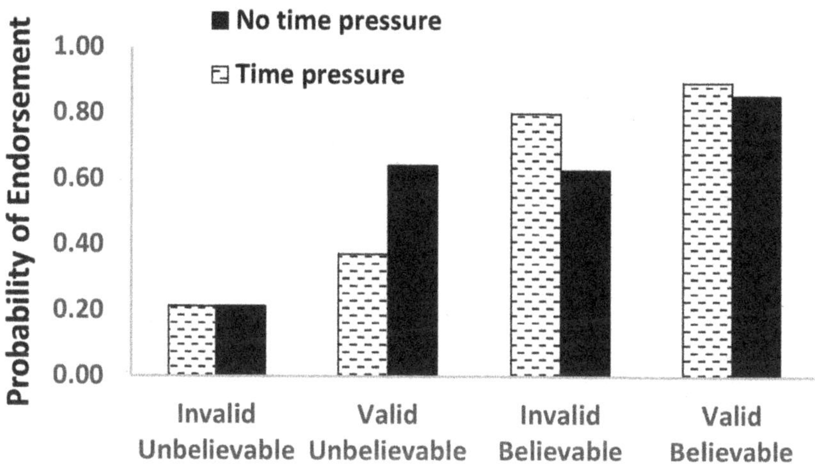

Figure 7.1
Results from the experiment by Evans and Curtis-Holmes. Participants read the syllogisms in table 7.2 and for each one decided whether the conclusion necessarily followed from the premises. In this graph, the vertical axis shows the average rate at which participants endorsed the conclusion of each type of syllogism as valid. The syllogisms were either valid or invalid, and their conclusions either believable or unbelievable. When decisions were made under no time pressure (solid bars), both validity and believability influenced choices. Under time pressure (hatched bars), in contrast, validity had only a very small effect on decisions and believability a much larger effect. At face value, this pattern seems to suggest that there are two different mental processes or systems that contribute to logical reasoning. The slow, logical system 2 dominates when ample time is available, but this system makes a smaller contribution under time pressure. The fast system 1, which promotes the intuition that believable conclusions are likely to be valid, dominates under time pressure.

and third—the invalid ones). In addition, believable arguments (third and fourth bars from left) were judged stronger on average than less believable ones (first and second bars).

Strikingly, time pressure had opposite effects on these influences. Now participants' judgments were less affected by validity but more affected by believability. This can be seen in the hatched bars in the figure. Validity had only a very small effect on the endorsement rate (the heights of the first and second hatched bars are similar, as are the third and fourth despite differing in validity), while believable syllogisms (whether valid or invalid) were much more likely to be judged acceptable than unbelievable ones (the third and fourth hatched bars—believable—are much higher than the first and second—unbelievable).[5]

The line from this pattern of results to the system 1/system 2 distinction is a straightforward one and is illustrated in the upper (dual-process account) section of figure 7.2. When there is no time pressure, the slow, logical system 2 is able to evaluate the strengths of the arguments using rational principles to distinguish valid from invalid deductions. System 1 also gets in on the act, creating a modest draw toward the more believable conclusions. Under time pressure, on the other hand, system 2 cannot operate as it normally would; hence, logical validity determines participants' judgments to a lesser extent, and the believability bias is enlarged as system 1 is "unleashed" to strongly influence responses.

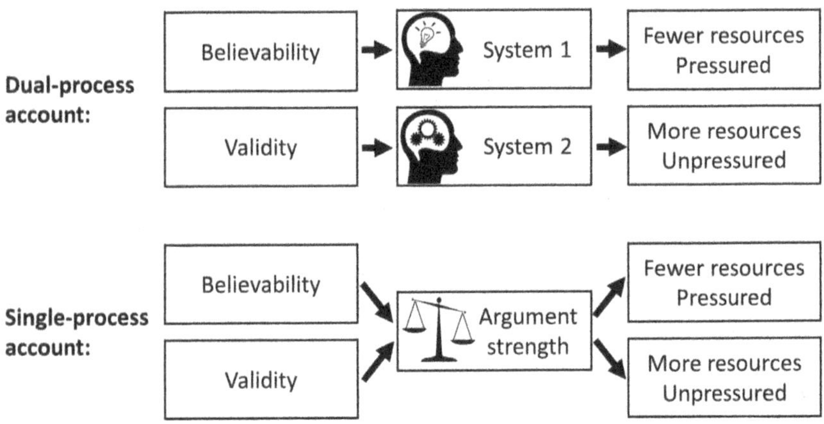

Figure 7.2
Two contrasting theories for explaining the findings of Evans and Curtis-Holmes. The dual-process account assumes the existence of a fast, automatic, and unconscious system 1 and a slow, conscious, and logical system 2. System 1, which dominates under time pressure, is intuitive and hence leads to choices that are heavily influenced by the believability of the conclusion of the syllogism. System 2 plays a large role when ample time is available, is logical, and leads to choices that are influenced by validity. The contrasting single-process account assumes that there is a single continuum of argument strength whereby valid syllogisms with believable conclusions are strongest and invalid ones with unbelievable conclusions are weakest. Both believability and validity hence contribute to this single measure of argument strength. Different mappings of argument strength onto behavior are responsible for the dominance of believability under time pressure and the greater influence of validity when time is available.

Observation and Inference

What is wrong with this account? A fundamental problem in psychology is the challenge of inferring the nature of the mind from outward behavior. Psychology is often defined as the science of behavior, a concept that emphasizes that observable behavior is to a large extent all we have to go on if we want to build theories about mental processes. In some aspects of perception and action, we might be able to measure activity in specific parts of the brain that are strongly associated with the content of our thoughts. For example, we can identify very specific brain regions associated with the perception of different colors or sounds and can use this knowledge to read the mind. From information about brain activity, it is possible to infer what underlying mental state the person is in, such as looking at a red or green patch. But these cases are the exception rather than the rule. They relate strongly to what we might think of as input or output processes in the brain. When it comes to thought more generally, we have to infer mental states. We cannot directly observe mental processes associated with whether some ongoing mental calculation is correct or how close a person is to reaching a difficult decision.

So the underlying states we refer to in explaining behavior, such as beliefs, desires, attitudes, and feelings, are hypothetical entities. Much of psychology and cognitive science is dedicated to the process and methods by which these "latent" states can be characterized from their effects on behavior, but there will inevitably be uncertainty about their relationship to behavior. Consider the example of two different ways of measuring knowledge, via speed and accuracy. One way of measuring individuals' general knowledge would be to give them unlimited time to answer a set of, say, fifty trivia questions. This method emphasizes accuracy of knowledge, with speed being irrelevant. Another way would be to give everyone a limited amount of time and measure how many questions they're able to answer correctly in that time. This method places more emphasis on speed of accessing one's knowledge. We might find somewhat different patterns from these measures. Person A might score much higher than B when unlimited time is available, but they might perform the same under time pressure. Person C might score better than D under time pressure but perform the same with unlimited time. It is, in short, difficult for us to know very much about the precise way in which underlying mental states map onto observable behavior.

For our psychological theories to be useful and falsifiable, there have to be some constraints on these mappings. A very strong constraint would be that they are linearly related, such that an X percent increase in the latent state (such as the amount of general knowledge) always maps onto a Y percent increase in our measurement of that knowledge. Under such a linearity assumption, this X-Y relationship would apply whether the person's general knowledge is poor or excellent. A much weaker constraint would be *monotonicity*, the milder assumption that if the latent state increases in strength, the observable behavior should either remain constant or increase but never decrease. This assumption captures the commonsense idea that as a person becomes more knowledgeable, she should always score at least as well, and never worse, on a measure of that knowledge.

The implausibility of linear mappings between mental states and behavior can be seen very easily by considering all-or-none actions. Think about your decision to take an umbrella with you when you leave home in the morning. You presumably have some strength of belief, between 0 percent and 100 percent, that it will rain today. Perhaps you've looked at the weather forecast or seen gray clouds out of the window. If we accept that your underlying belief lies on a continuous scale, then the linearity assumption says that your behavior—whether to take an umbrella—must also vary continuously in strength. But this is impossible: you either take an umbrella or you don't. In this case, it's obvious that the mapping is a step function. Between complete belief that it won't rain and some lower level of belief strength, say 30 percent, you choose not to take an umbrella, but as soon as your belief exceeds that level, you take an umbrella. Between that 30 percent level and certainty (100 percent), there is no behavioral indicator of your strength of belief. Your observable behavior (taking an umbrella) is identical whether the belief is 30 percent or 60 percent or 100 percent.

Latent States and the Implausibility of Dual Systems

How does this relate to the two-systems debate? The answer is that many of the most apparently convincing arguments for dual systems, including the dual-systems interpretation of Evans's findings, rely on extremely strong, and probably implausible, assumptions about mappings between latent mental states and behavior. If there is a linear mapping from underlying belief strength to endorsement rates, then Evans's results indeed suggest that two distinct systems or processes contribute to people's decisions. This can be seen most clearly in figure 7.1 by contrasting endorsement rates to the valid but unbelievable (VU) and invalid but believable (IB) syllogisms. Under time pressure, the IB syllogisms (third hatched bar from left) were more than twice as likely to be judged valid as the VU ones (second hatched bar), and hence to explain this difference we have to assume that the underlying belief strengths of these syllogisms differ greatly. However, if latent belief strengths map linearly onto behavior, this difference in the underlying strengths of the VU and IB syllogisms should also be evident when there is no time pressure, yet this is clearly not the case, as the figure shows (the second and third solid bars are almost identical in height). So it would be plausible to conclude that whatever is the latent process that is responsible for the difference in endorsement rates to these items, it is something that influences choices under time pressure but not when there is no time pressure. This is in effect what the dual-systems account proposes, with system 1 being the label for this process.

As soon as we relax the linear mapping assumption, this line of argument becomes much less compelling. Let us assume instead that the mapping from latent belief strength to endorsements has the form shown in figure 7.3, with valid and believable (VB) syllogisms having greatest strength, followed by IB, VU, and IU items in that order and with slightly different mappings when time pressure is or isn't applied. Both of these functions are monotonic in the sense that as strength increases across the four item types, endorsement either increases or at least remains steady. Crucially, an increase in strength is never accompanied by a decrease in endorsement rates.

With these mappings, Evans's data are easily explained. When there is no time pressure, validity is a major influence in that going from an invalid to a valid syllogism (from IU to VU, or from IB to VB) results in a large increase in predicted endorsements. At the same time, believability is also important: going from IU to IB or from VU to VB results in an increase in endorsements. So long as the syllogisms are ordered in the right way, the mapping is monotonic. In the contrasting case where decisions are made under time pressure, the mapping yields a different pattern in which believability is the dominant factor. Going from an unbelievable to a believable syllogism (from IU to IB or from VU to VB) results in a large increase in predicted

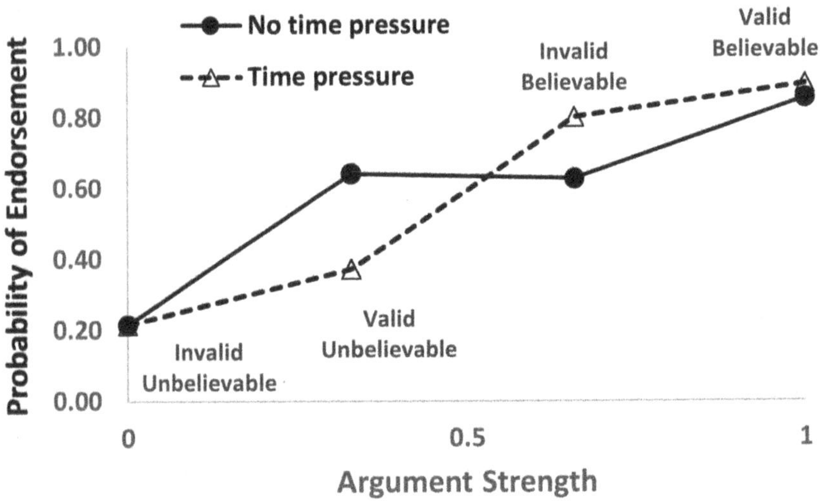

Figure 7.3
Hypothetical mappings from belief strength to behavior across the four types of syllogism that Evans and Curtis-Holmes used. The data are identical to those in figure 7.1, but the horizontal axis now orders the four syllogism types according to their hypothetical latent strength, with valid and believable (VB) syllogisms having greatest strength, followed by invalid believable (IB), valid unbelievable (VU), and invalid unbelievable (IU) ones in that order. The vertical axis indicates the probability that each type of syllogism is endorsed as valid. If slightly different functions are assumed for the time pressure and no time pressure conditions, Evans and Curtis-Holmes's results are perfectly reproduced, as indicated by the symbols on each line, which exactly mirror the results shown in the earlier figure. Crucially, both of the lines are monotonic in the sense that they always increase or stay flat as argument strength increases, and they never go down.

endorsements, whereas validity has a smaller impact, in that going from IU to VU or from IB to VB results in only a small increase in endorsements. Again, as long as the syllogisms are ordered in the right way, the mapping is monotonic.

In short, the results of this influential experiment only provide support for the dual-systems theory if a very strong assumption is made about the mapping from the underlying latent argument strengths to behavior—namely, that it is linear. But we have no evidence that it has this form. On the much weaker and more plausible assumption that this mapping is very unlikely to be perfectly linear, nothing in the results requires the postulation of two separate systems.

The astute reader may be wondering how this single-system account could ever be falsified. The answer is that any pattern of findings that cannot be reproduced by monotonic mappings would seriously question it. Suppose that we obtained data revealing that the different syllogisms cannot be ordered monotonically in the two conditions, time pressure and no time pressure, by any increasing measure of syllogism strength. For instance, imagine that Evans had found that endorsement rates to the VU syllogisms were greater than to the IB ones under conditions of no time pressure. This would mean that there is one ordering of endorsement rates (IB > VU) under time pressure but a different ordering (VU > IB) under no time pressure. There is no ordering of these items by latent strength that can generate such a contradictory pattern, and hence the single system account would be demonstrably inadequate, even with its very weak monotonicity assumption. However, extensive investigations of very many experiments have failed to throw up convincing evidence of such patterns.[6]

This reinterpretation of how people solve logical riddles casts serious doubt over the plausibility of dual-systems accounts. But are some simpler mental processes truly automatic in the sense of being completely outside our voluntary control?

Stroop: A Paradigmatic Automatic Effect?

In 1935 a young American scholar by the name of John Ridley Stroop published a paper entitled "Studies of Interference in Serial Verbal Reactions" that was to have a profound and enduring impact on psychology.[7] The phenomenon described in that paper now bears the author's name and is known to graduates of psychology programs the world over: the Stroop effect. The effect, which we will come to in a moment, has become synonymous with the idea of conflict between automatic and effortful, controlled processes. It is often taken as prima facie evidence of the mind's internal struggle between dual monolithic systems of cognition: one unconscious and outside voluntary control (system 1), the other conscious and controllable (system 2).

Perhaps surprisingly, neither the words *automatic* nor *controlled* appear anywhere in Stroop's original paper, and there is certainly no mention of dual systems, or consciousness. It is of course impossible to know what J. R. Stroop would have thought of his effect becoming so influential in the development of dual-systems thinking—he died in 1973, long before the current zeitgeist for dichotomies—but there are some clues. Thirty years after his original paper was published, he claimed to have no interest in the task he had created.[8] The "Stroop Scholar" Colin MacLeod recounts that Stroop's son told him that his father's psychological research was "insignificant to his Bible-oriented life and teaching."[9] Indeed, Stroop became a biblical scholar of some standing, preferring that calling to studying the apparent inherent conflict between mental processes.

The basic Stroop phenomenon is incredibly easy to demonstrate. Consider the examples presented in figure 7.4. For each panel, the goal is to read out loud the color of the ink that the word (panels a and b) or single letter (panels c and d) is written in. (For the purposes of this grayscale illustration here, you will have to imagine that the dark gray ink is blue and the light gray ink is yellow). Starting with panel a, you'll find it is pretty easy (and quick) to say "blue" and "yellow," but for panel b, it is trickier. There

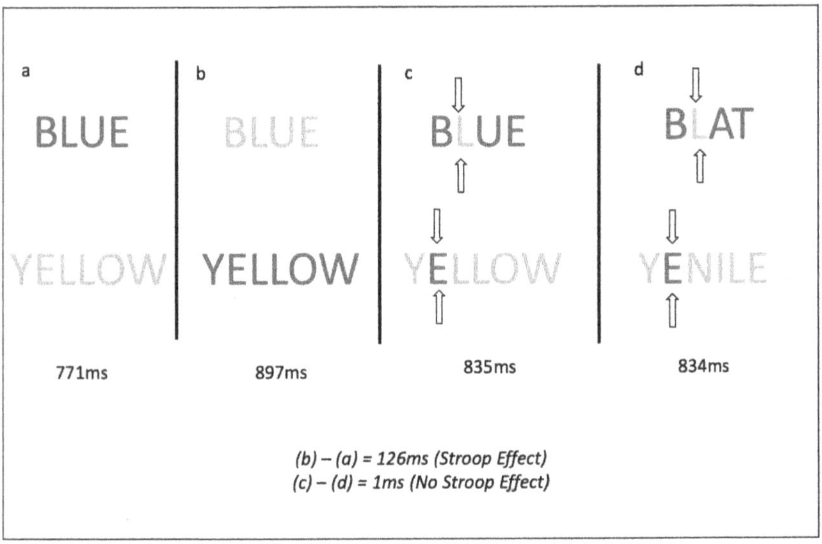

Figure 7.4
A demonstration of the Stroop effect and its disappearance. The task in each panel is to name the color of the ink that the word or cued letter is printed in. The difference in reaction times between panels a and b is often taken as evidence for the automaticity of word reading. However, finding that reaction times are identical in panels c and d casts doubt on this interpretation. (Note you need to imagine that the dark gray ink is printed in blue and the light gray in yellow.)

is a strong temptation to say "blue" instead of "yellow" for the first word and "yellow" instead of "blue" for the second. This difference in the ease of naming the color between panels a and b is the standard Stroop effect. In this example, on average, participants took 126 milliseconds longer to identify the correct color in the incongruent (panel b) compared to the congruent (panel a) case.[10] (An interesting, if admittedly arcane aside here, is that although the comparison of naming times in the congruent and incongruent versions is the standard Stroop effect, Stroop himself never actually gave participants the "congruent" version. Instead, he compared how long it took participants to read words versus name colors given lists of incongruent words like those in panel b.)

So why the difference in reaction times? The standard story is that the incongruent version takes longer because word recognition is automatic: lexical (determining the pronunciation) and semantic (determining the meaning) analyses of words are inevitably triggered by the presentation of a word. We cannot help but read the word and access its meaning, "BLUE," in the top line of panel b, and this triggers a voice inside our heads saying, "Say blue! Say blue!" Our controlled, conscious voluntary intention to say "yellow" is at the mercy of the all-powerful automatic process. But is this simple dichotomy, this conflict between systems, the whole story?

Consider now panels c and d of figure 7.4. Here, the task is to read out the color of the letter that is pointed to by the white arrows. Give it a try. In panel c, the color of the cued letter is incongruent with the whole word (the letter L is yellow, but the word says blue); in panel d, however, neutral "nonwords" have replaced the color words, but again a single letter is colored. Researchers often use these kinds of nonwords as controls to isolate particular aspects of visual and cognitive processing. The complete absence of a difference in reaction times between panels c and d is telling. If the mere presentation of a word inevitably and automatically triggered recognition and the processing of meaning, then we'd expect color naming to be slower in panel c than in panel d. Why? Because the word "blue" has a meaning that if processed automatically should interfere with our ability to say "yellow" in response to the cued letter L. In contrast, "BLAT" is meaningless and thus should not cause us any problems when naming the yellow L. The same goes for the blue E in "YELLOW" and "YENILE." The fact that we do not see any evidence for interference—the reaction times are almost identical—strongly suggests that word reading was not automatically triggered via the setup in panel c.

A better explanation for what happens in panels c and d is that we are able to exert some control over the deployment of our attention. Specifically, the arrows cue us to focus on a particular part of the word, thereby suppressing the word recognition processes that would otherwise lead to interference and slow naming speed. The words "BLUE" and "YELLOW" are right in front of our eyes to exactly the same extent in panels b and c, but we don't read them to the same extent: in the panel b setup, we do read them and suffer from Stroop interference, whereas in the panel c setup, we don't. So word reading cannot be an automatic process.

In fact, all kinds of variants of the Stroop task provide evidence that is more consistent with the idea that interference effects are graded rather than binary. Simply adjusting the ratio of congruent and incongruent words in a list affects the size of the Stroop effect. When congruent words dominate, the interference effects are larger because the overall context of the experiment encourages reading the word (rather than naming the ink). If word recognition were purely automatic, the ratio should not matter.[11]

Pragmatism versus Accuracy

Dual-systems thinking is deeply ingrained in contemporary discussion about the mind, as our earlier illustrations exemplify. How should we regard this framework in light of the kinds of counter-evidence we have just described? Daniel Kahneman stated in his influential book *Thinking, Fast and Slow*, "I must make it absolutely clear that [system 1 and system 2] are fictitious characters" that nonetheless, in his view, serve a useful purpose.[12] That pragmatic purpose is to make it easier to communicate complex ideas about the mind. The logic here is that it is comparatively easy for the general public to comprehend the notion that the mind can bring two quite distinct modes or systems of thought to bear in tackling a decision problem. Those two systems can be contrasted via a series of binary features such as intuitive versus rational, fast versus slow, automatic versus deliberate, hot (emotional) versus cold (unemotional), and so on (see table 7.1).

Such a framework can of course be helpful so long as it is accurate and there is reasonable consensus about the features of the two systems. But it will become a positive hindrance if the features are poorly defined and, worse still, if it turns out to be unsupported by the evidence. Scientists have sacrificed forests in proliferating variant dual-system models, each with its own characteristics. And in devising these models, which more and more resemble the epicycles devised by Greek astronomers to rescue the theory that the Earth revolves around the sun, their contact with hard facts becomes more and more distant. It is very rare to see proponents of the dual-systems approach provide explicit details of the predictions their models make that could be subjected to experimental tests.

Each of the individual dimensions mentioned above, such as automatic versus deliberate and emotional versus unemotional, is of great importance in understanding the mind, but the dual-systems approach provides very little explanation of why these dimensions should align. Why can we have automatic, emotional, or deliberate unemotional thinking but not deliberate emotional thinking? Our view is that behavioral scientists should first and foremost seek to understand the ways in which each of these dimensions is relevant to the mind. When, in the far distant future, we have a near-comprehensive grasp of them, then we can ask whether the dimensions tend to align or not. In our present state of knowledge, however, the degree of alignment assumed by dual-process accounts is little more than an untested assumption.[13]

There is also very little hard evidence that modes of thinking are binary rather than continuous. Both of the experimental effects we have discussed in detail in this chapter, logical reasoning and the Stroop effect, are better thought of from a continuous perspective. In the case of logical argument, we saw that the effects of validity and believability can be understood in terms of continuous argument strength. We also saw that whether a color word causes Stroop interference is a graded phenomenon. Many researchers now think of automaticity as a graded property, with different mental tasks placing demands on attention to varying degrees.[14]

To summarize, dual-systems thinking has become widely adopted not only among psychologists but more widely in debates about economic behavior, health, and public policy. This viewpoint may serve some useful communicative functions, such as conveying the important point that not all human decision making is based on logical or rational principles.[15] However, beyond this pragmatic function, the framework has a number of other implications, not all of them positive. It encourages binary thinking in places where it may not be appropriate, and it invites the view—for which there is very little evidence—that mental processes fall into clusters of aligned features.

This tension provides a segue into the second part of the book where we start to think in more depth about why the shaky theoretical foundations on which many claims about the unconscious mind are based have come to be ignored, forgotten about, or simply not known. (Oh, and in case you are still wondering, the answer to the bat and ball question is five cents.)

II Toward a True Understanding of Mind and Behavior

8 Feeling the Future; Precipitating a Crisis

In part I, we dove into the murky depths to search for evidence of a smart unconscious. We discovered that many high-profile examples of unconscious influences evaporate once scrutinized, or at least admit alternative and often more plausible explanations. In many ways, it seems that the very notion of unconscious thought is misguided. With this new perspective on the close connection between awareness and behavior established, we now turn to the equally puzzling question of how we, as a discipline, but also as a society, got to this point. How did we become hoodwinked into believing that our unconscious mind has a hold on our behavior?

Anatomy of a Train Wreck

In early October 2012, Ed Yong, a staff writer for the scientific journal *Nature*, published a short article with a long-lasting impact.[1] The article discussed an email written by Daniel Kahneman, a Nobel laureate and perhaps the world's best-known psychologist, whom we read about in chapters 3 and 7. The email described Kahneman's growing concerns about an area of psychology known as "priming." As we saw in chapter 3, priming is the study of how subtle cues can apparently unconsciously influence our thoughts and behavior. For example, a person might walk more slowly down a corridor when thinking about elderly people or find cartoons funnier while "smiling" when holding a pen in her mouth.

In the email—intended for colleagues but leaked to *Nature*—Kahneman wrote provocatively of a "train wreck looming" because of doubts about the replicability of many priming effects. Kahneman's concern followed revelations of fraudulent conduct by some researchers. The pressure to publish or perhaps the desire for fame had led some to simply invent data. Less dramatic but equally troubling, many researchers were unable to reproduce prominent, eye-catching findings and the bulging file drawers in which these "failures to replicate" had been hidden were finally spilling open.

The implication of these events for the "integrity of psychological research" played strongly on Kahneman's mind. He advised colleagues to "collectively do something about this mess" and suggested methods for improving the replicability of research. In this brief chapter we provide the context behind the looming train wreck. We set the scene for chapters 9 and 10, which explore in more detail the nature of the research practices that precipitated the crisis in confidence and the scientific ecosystem that encourages behaviors that are at variance with the pure pursuit of truth.[2]

* * *

It was the question Diederik Stapel had been dreading for years: "Diederik, I have to ask you: have you been faking your data?" Stapel, a professor of social psychology at the University of Tilburg at the time, was sitting in the living room of his friend, Maarten, the chair of the department. The question was direct and unexpected. Stapel's first reaction was to deny the accusation flat out. Faking data was, as Stapel would write in his memoir *Derailment* a few years later, "professional suicide."[3] An admission of guilt would end his extremely successful career. Instead he played it down. What evidence did Maarten have? Were there specific details, or was it just idle gossip at conferences?

Maarten had been taken aside after dinner at a recent conference by some young researchers who had collaborated with Stapel. What they told Maarten was deeply unsettling. The junior colleagues claimed that no one knew where Stapel was getting his data from. They feared that he'd simply been making up numbers. Stapel's initial attempts to mollify Maarten appeared to work. The rest of the evening was genial—discussions about interesting presentations Maarten had seen at the conference—but Stapel knew that his "big, fat, outright lies" were now likely to become very public.

To get a sense of the size of these lies, consider the following example of one of Stapel's studies that turned out to have been rather economical with the truth. Imagine that you are walking down the street and are stopped by a person holding a clipboard. The interviewer asks you about your thoughts and feelings toward particular minority groups in society (perhaps Muslims). The goal, although not stated explicitly, is to measure your tendency to stereotype people. After answering the questions, the interviewer offers you a payment of 5 euros but also asks whether you'd be willing to donate a portion of the money to a charity that helps immigrants and homeless people. You give back a few euros, you are thanked, and you head off down the street. This is a simple interaction, but is something more complex going on? Collecting data in the field like this is a common approach in many areas of psychology, especially those concerned with social interaction. There is a strong desire to get outside the confines of the sterile laboratory and into the messy "real world." In the study just described, it was literally a need for mess that pushed Stapel out on to the streets. He wanted to explore the broken windows theory—the idea that poor physical environments (dilapidated buildings, broken windows) beget poor social environments. His extension to this idea was that messy and rundown physical environments lead people to rely more on stereotypes and other forms of prejudice because stereotyping allows people to create structure in their mental world. In Stapel's words, "Stereotyping is a mental cleaning device that helps people to cope with physical chaos."[4]

What does this have to do with being stopped on the street and asked about your views on different groups in society? The key manipulation in the "coping with chaos" study was the presence or absence of messy street features. For one group of participants, the interviewer was standing in a spot with broken tiles on the pavement, a poorly parked car with its windows open, and an abandoned bicycle. For the other group, tested in exactly the same place on another occasion, none of these features were present: the street was neat and tidy. Stapel predicted that when people were asked about their thoughts and feelings in the presence of "chaos," they would show more stereotyping. That is, they would accentuate the differences between us (the Caucasian participants) and them (the various minority groups listed in the survey) to compensate for the disordered physical environment. Moreover, he predicted that those faced with chaos would donate more of their participation fee to the Money for Minorities charity than those on the ordered street.

This pattern of results is exactly what Stapel found. Except he didn't. Because none of it happened. No interviewer, no passersby, no broken bicycles. No (real) data. And yet the paper on these results was reported in one of the world's most prestigious scientific outlets, the journal *Science*. How could this have happened?

Perhaps the simplest answer to this question is that Stapel had not wanted inconvenient data to get in the way of a good story. In fact, he had not wanted *any* data to get in the way. The idea that chaos might lead to stereotyping had come to Stapel several years before the faked street study. He had done some initial laboratory-based experiments in which he showed participants photos of disorderly and orderly scenes (walls with or without graffiti) and then had them rate their feelings toward different social groups. He had found small differences consistent with his hypothesis—the people shown the disorderly scenes tended to display more prejudice—but the differences were weak and transient. After a few more failed attempts to find the effect, he gave up. Eventually, though, the lure of a good idea and a good story to go with it overpowered Stapel's need for data.

For the kinds of experiments reported in the coping-with-chaos paper, faking data is relatively straightforward. Each (hypothetical) participant provides some numbers on a few rating scales indicating how much they tend to stereotype; these numbers can then be combined to produce average or mean ratings for the different conditions (chaotic versus ordered street scenes) and subjected to statistical analysis. Stapel describes the process in some detail in his memoir:

> I would . . . make a careful list of all the results and effects I needed to create for the experiment I was doing. Neat tables with the results I expected based on extensive reading, theorizing, and thinking. Simple, elegant, comprehensible. Next I started to enter the data, column by column, row by row. I tried to imagine how the participants' answers to my questionnaire would look. What were some reasonable answers that might be expected? 3, 4, 6, 7, 8, 4, 5, 3, 5, 6, 7, 8, 5, 4, 3, 3, 2. When I'd input all the data, I ran some quick preliminary analyses. Often these didn't show what I was expecting, so I went back to the table of data to change a few things. 4, 6, 7, 5, 4, 7, 8, 2, 4, 4, 6, 5, 6, 7, 8, 5, 4. And so on, until the analyses provided the results I was looking for. That is, until the data showed what was logical, and therefore true.[5]

The brazenness of Stapel's approach is chilling. His ability to have gotten away with it for so long had emboldened him. If we are to believe his memoir, faking the data made him nauseous and terrified him but he was addicted—addicted to the success and fame that the next big (fake) discovery (and publication) would bring him. Stapel had been playing a dangerous game, but he'd been playing it carefully. The results he created were simple and comprehensible—they made a good story—and they fit, more or less, with established theories. Thus, although some of the statistical results might have appeared too good to be true, the claims being made were not *too* extraordinary. We can make sense of the idea that messy surroundings are unsettling and that we might vent some frustration at the chaos by distancing ourselves from those we perceive as responsible.

Achieving this high-wire act of balancing plausible yet sufficiently surprising findings had opened the door to a slew of high impact publications for Stapel. However, the Levelt Report—an investigation into Stapel's fraudulent activity—concluded that in addition to his own dishonesty, there had also been serious shortcomings in the publication process and a widespread failure of scientific criticism when it came to his work.[6] The report established that the peer-review process (where other researchers get to evaluate work before it can be published) had often been "strongly in favor of telling an interesting, elegant, concise and compelling story, possibly at the expense of the necessary scientific diligence." There was, it seems, an uncritical acceptance of findings that "felt right" despite a lack of clear evidence, let alone reliable replication.

Although Stapel had created a raft of false-positive results—or simply false results–he had been careful not to make his claims too extraordinary. He'd prided himself on keeping the stories of his faked data simple, elegant, and within the bounds of existing theory. But sometimes the pendulum swings the other way. The data may be genuine but the theoretical claims made on their behalf are simply off the map. And when the claims are extraordinary, as the astronomer Carl Sagan was fond of saying, the evidence needs to be extraordinary too. Precognition anyone?

Time-Traveling Porn

"Science has finally discovered time-traveling porn," declared Stephen Colbert on his hugely popular US TV show *The Colbert Report* in January 2011. The topic: the claim by Cornell social psychologist Daryl Bem to have found evidence that participants in his experiments could "feel the future"—specifically that they knew, in advance, what picture was about to be displayed on a computer screen in front of them. But not any old picture; it only worked for erotic images. In discussing his results, Bem quoted from the celebrated quantum physicist Richard Feynman: "Do not keep saying to yourself . . . 'But how can it be like that?' because you will get . . . into a blind alley from which nobody has yet escaped. Nobody knows how it can be like that." Quite.[7]

The experiment itself was very simple.[8] Cornell University undergraduates sat in front of a computer monitor displaying images of two curtains side by side. They were told that behind one curtain was a blank wall and behind the other was a picture. All they needed to do was "click on the curtain that you feel has the picture behind it." Once clicked, the curtain was drawn back to reveal either the picture or the wall. The procedure was repeated for thirty-six trials. The position of the picture, relative to the wall, was randomized on each trial, as was the type of picture displayed. On some trials, the pictures were erotic ("couples engaged in nonviolent but explicit consensual acts"), on some they were negative images like snakes and spiders, and on others they were positive and romantic but not erotic, such as a bride and groom kissing at their wedding. A final type was simply neutral (landscapes).

With a fifty-fifty chance of picking the curtain covering the image, one would expect people to be 50 percent accurate—that is, at chance—across the thirty-six trials of the experiment. The extraordinary finding was that the hit rate (choosing the correct curtain) for the erotic pictures was 53.1 percent. This might not seem like much above the expected 50 percent, but it was statistically significant (an issue to which we will return), which is to say meaningfully greater than 50 percent. Moreover, as Bem pointed out to Stephen Colbert, although it seems small, it is in fact similar to the margin by which Obama defeated McCain in the 2008 presidential election; it is also about the same margin the house has over the punter in casinos. Hit rates for the nonerotic pictures—whether neutral, positive, negative, or romantic—all fell very close to 50 percent and did not pass the statistical test for being a "real" result. So people can feel the future, but only if it is porn! Stephen Colbert, along with large chunks of the media, had a field day with this finding.

Within academic circles, the reaction was one of puzzlement but also deep concern. Here was an article published in one of the top journals in social psychology reporting apparent evidence undermining a fundamental belief in the direction of causation. We all know that a cause must precede an effect: you cannot hear the bell ringing until someone strikes the bell. But Bem wanted us to believe that the content of an unseen image can have a retroactive influence on our choice: "The participant is, in fact, accessing information yet to be determined in the future, implying that the direction of the causal arrow has been reversed."[9] The yet-to-be-determined aspect is a subtle but crucial feature of the experimental design. In the experiments, participants made a selection *and then* the computer used a special algorithm to randomly determine which picture would be displayed behind the selected curtain. Thus, the participants were not displaying clairvoyance; they were showing *precognition*. This is knowledge of a future event that could not otherwise be anticipated through any known inferential process.

Psychology has a long tradition of interest in psi phenomena: anomalous processes of information or energy transfer that are currently unexplained in terms of known physical or biological mechanisms. This fascination seems in part to be driven by the widespread acceptance among the public that something like psi exists. For example, according to a US CBS Newspoll, 57 percent of Americans believe in extrasensory perception, and other surveys show that over 40 percent of people believe in telepathy.[10] *If so many people believe in it, then perhaps there is something worth investigating* appears to be the logic that Bem and his fellow parapsychologists relied on. But the old saying that the plural of anecdotes is not anecdata has long been recognized as relevant here. Joseph Jastrow, one of the founders of modern psychology, wrote a book, *Fact and Fable in Psychology*, way back in 1901 that was largely concerned with separating what he saw as the legitimate inquiry of the then relatively new discipline of psychology (the facts) from the vagaries of "psychical" research (the fables). He suggested:

> If the problems of psychical research, or that portion of the problems in which investigation seems profitable, are ever to be illuminated and exhibited in an intelligible form, it will only come about when they are investigated by the same methods and in the same spirit as are other psychological problems.[11]

Arguably, by bringing phenomena like precognition into the laboratory and performing simple and potentially easily replicable experiments, Bem was following this advice. He went beyond anecdotal reports and attempted to use statistical methods to demonstrate reliable—and yet inexplicable—effects in his data. But again, some prescient words from Jastrow are worth heeding here:

> Data cannot claim serious attention before they are strong in their validity, and extensive in their scope and consistently significant in their structure; then, and not before, are they ready for the crucible of scientific logic, from which they may

> or may not emerge as standard metal, to be stamped and circulated as accepted coin of the realm.[12]

As Bem was about to discover, many in the discipline were not quite ready to accept his findings as legal tender. E.-J. Wagenmakers and colleagues from the University of Amsterdam were quick to critique Bem's research.[13] They urged fellow psychologists to reconsider the ways in which they analyze data and to be clear about the differences between *exploratory* and *confirmatory* research. Exploration is fine—it is often the way we make new discoveries; noticing the anomalous pattern in the data (or the mold in the petri dish) can lead to significant breakthroughs. But novel hypotheses discovered on an exploratory trawl through the data must be tested in new, confirmatory experiments. A single surprising finding is just that—a one-off that might have occurred simply by chance. To have confidence that a finding should be newly minted and circulated to the community, it needs to be replicated. And the more surprising the finding, the more urgent the need for replications.

Wagenmakers and colleagues pointed to several instances of what they perceived to be exploratory practices in Bem's feeling-the-future paper. Why, for example, was precognition only found for erotic and not neutral, negative, or positive pictures? Was this an a priori prediction or something discovered by slicing and dicing the data after collection? Why in some experiments were anomalous influences found only for women and not men? According to Bem, the psi literature does not show systematic sex differences in psi ability, so why test for gender unless to squeeze out something—anything—interesting? The problem with this exploration is that each time the data are examined in a different way, the possibility of finding a false-positive result—a result that looks genuine, but is in fact spurious—increases. Researchers know these as type 1 errors.

The principal guard against such errors is to use statistics—essentially methods for inferring whether observed patterns in data are real or occurred by chance. The most commonly used technique in psychology (and many other disciplines) is null hypothesis significance testing (NHST). This is what Bem (and Stapel) used. The basic idea is to test whether the data provide enough evidence to reject the hypothesis that there is no impact of a manipulation on behavior—the so-called *null hypothesis*. Rejecting the null when in reality there is no evidence for a difference in the data is the definition of a type 1 error. So why didn't the statistics that Bem used control for these kinds of errors? Why could he not be confident in rejecting the null hypothesis that the 53.1 percent hit rate for selection of the erotic images was a chance result? Wagenmakers and colleagues illustrate the problem of relying solely on NHST with the following example.

Imagine you have won $10 million in the state lottery—an extremely happy event, but one that might lead to jealousy in some friends and acquaintances. Perhaps this jealousy leads one acquaintance to accuse you of cheating: the probability of winning the lottery is very low so you must have cheated. We immediately see that this argument is not logical: the low probability by itself cannot be taken as evidence for cheating. The evidence becomes useful only if we compare it to an alternative hypothesis—with an even lower probability—that you were somehow able to obtain advanced knowledge of the winning numbers (perhaps you are a "precog"?).

The key point here is that the strength of evidence for a particular hypothesis—precognition exists—needs to be evaluated against a specific alternative hypothesis—precognition does not exist—rather than just a "nothing" or null hypothesis. Wagenmakers and colleagues argue that the best way to conduct such comparative hypothesis testing is by using a different statistical approach based on Bayesian methods. The details need not concern us, but the results should. When reanalyzed using these arguably more appropriate techniques, the evidence for precognition in all nine of Bem's experiments all but evaporated and indeed in three of the nine substantial evidence in favor of the nonexistence of precognition emerged. These reanalyses should give us considerable pause for thought. But even if the statistical arguments are rather abstract, there are other, more fundamental reasons for questioning Bem's conclusions.[14]

Klaus Fiedler, a social psychologist from the University of Heidelberg in Germany and long-term observer of the field, also found himself troubled by the publication of Bem's work. Writing with colleague Joachim Kruger from Brown University, they took Bem to task on the absence of any theoretical explanation of his findings and the confusion between the explanans (the argument used for explanation) and the explanadum (the event to be explained).[15] In essence, Fiedler and Kruger argued that Bem's article lacked any solid theoretical base. Although there were some allusions to quantum physics in the discussion of the results (and the appeal to Feynman), the bottom line was no real theory and no real explanation. Bem admitted as much to Colbert in his interview; when Colbert asked, "How is this working?" he replied, "We have no idea."

More than that, the explanation that Bem appeared to be seeking was of the wrong phenomenon. According to Fiedler and Kruger the explanadum and the explanans had been reversed. In Bem's experiments, the crucial finding was that participants chose the erotic pictures at a rate above chance. If we hang on to the (pretty-well-established) idea that antecedent conditions explain consequent events, then the event to be explained is the computer's bias to produce responses congruent with participants' choices. Or as Fiedler and Kruger write, if we think that the results are valid (which is of course questionable), then they "must be interpreted as owing to metaphysics in the computer's chips rather than precognition in the human brain."[16]

What this boils down to is a debate about whether random number generators used to determine the location of pictures was truly random or somehow correlated with the participants' preceding responses. While this might seem unlikely, the key question is whether it is less likely than a reversal in the laws of causal inference. Winning versus cheating on the lottery, random number errors versus psychic retroactive anomalous influences: it is all in the balance of probabilities. And as Sherlock Holmes famously advised, "When you have eliminated the impossible, whatever remains, however improbable, must be the truth."[17]

The Stapel and Bem cases and others like them have been very painful for the behavioral sciences, but they have also been useful. They have forced the discipline to introspect on its research practices and review the role of psychological theory in our explanations of mental life. It is hard to exaggerate the transformation in research methods and rigor that has occurred in the decade or so since Bem's research was published and Stapel's fraud was revealed, and our contention is that much of the evidence for unconscious influences on human behavior falls victim to these improved methods. In the next two chapters, we delve more deeply into the features of our research practices and the scientific ecosystem that got us into this mess (as Daniel Kahneman put it) and see yet again that weak claims about the powers of the unconscious play a crucial role. In the final chapter, we point to how a focus on transparency and strong theory can help get us out of the mess and back on track to reclaiming the science of the mind.

9 Research Biases

The achievements of science, technology, and medicine are all around us. Minor injuries or diseases that our ancestors would not have survived are routinely treated with surgical procedures, antibiotics, and other medicines. The science that makes a smartphone work is at the cutting edge of many fields: just the batteries themselves represent decades of research, culminating in the lithium-ion battery whose inventors won the 2019 Nobel Prize for chemistry. Our knowledge of the building blocks of nature derives from centuries of discoveries (such as the atom) and inventions (like the microscope).

Much of this science and technology is subject to immediate confirmation or refutation. We know that the science behind lithium-ion batteries is correct because they work. We know that ibuprofen is an analgesic because after taking it, our headache or other painful condition becomes less painful. The idea that a technology company could successfully market a new form of battery that in fact didn't work seems absurd. But despite these extraordinary successes, no one can seriously deny that substantial parts of science are in a state of crisis and are, to a large extent, broken. As soon as we move away from applications that provide immediate feedback about the veracity of the underlying science, we discover that there are no guarantees. The systematic way scientists work, collecting and analyzing data, and submitting their findings to journals for peer review, is not an infallible process for gradually accumulating correct knowledge about the world. Far from it. The scientific ecosystem, including grant-awarding panels that fund research, the research process itself, peer review, the institutional career progression and promotion mechanisms by which scientists are rewarded, and other components, is skewed in ways that lead to the generation of vast swathes of junk science. Much of this junk science concerns the unconscious.

This claim might seem extreme, but there are many reasons to believe it. In chapter 3, we described the phenomenon known as priming, where our perceptions or judgments or decisions can be influenced by seemingly unrelated events. Reading the word *bread* can induce a carryover effect and make us faster to subsequently read the word *butter*, and identifying a dalmatian dog in a black-and-white image (figure 3.1) can make it easier and faster for us to see the same dog in the image months or even years later. A rather more surprising form of priming was first reported and given a catchy title ("money priming") in 2006 by Kathleen Vohs and her colleagues and studied in dozens of subsequent reports (a recent review of this literature identified—incredibly—nearly 250 experiments, most of which found the effect).[1] The typical observation is the apparent modification of people's behavior on a variety of measures following exposure to images of money or tasks that involve subtle activation of the concept of money. For instance, the original study claimed that money priming causes people to work harder on difficult tasks and to become less willing to help others.[2] If this is true, the idea that workers can be nudged to exert more effort simply by subtle reminders of money is a distinctly nontrivial discovery, as is the finding that playing with coins makes children more selfish. Related research has claimed that subtle reminders of achievement, such as a photograph of a woman winning a race, can have a similar effect. In one study, for example, showing this photograph to employees in a fundraising call center increased the amount of money they raised.[3]

Later research claimed that viewing images linked to money (such as pictures of $100 bills) made people more likely to endorse free market values and social inequality.[4] They became more likely to agree with statements such as, "Some groups of people are simply inferior to others," for instance. Priming effects of this sort are explained by the unconscious activation of concepts (the mental idea of money in this case) and other closely related concepts.[5] We discuss money priming at some length in this and the next chapter for several reasons. Priming effects have been extremely influential in recent claims about the power of the unconscious mind and so deserve close scrutiny. Money priming, one of the most straightforward and intensively studied priming effects, is a veritable petri dish for considering the many biases, and the remedies for those biases, that have been identified in behavioral research over the past few years. As such, it stands as a revealing case study.

It seems extraordinary to imagine that something like money priming, documented time and time again in peer-reviewed journal articles, could be anything other than a true effect. Of course, there will always be limits to any phenomenon, and one would expect some money priming experiments to be failures. If the time interval between the money prime and the behavioral measure is too long, or if the prime is imperceptible or too subtle, surely the effect will become too diluted to be detectable. But this is not the problem here. Instead, the entire edifice of research on money priming is built on sand. There is (almost certainly) no genuine money priming effect.

Several lines of evidence point to this conclusion. After the initial flurry of studies on the phenomenon, researchers eventually undertook several very large efforts to replicate the early findings, and these efforts proved to be strikingly unsuccessful (we discuss these in detail in the next chapter). As doubts about this variety of priming began to accumulate, more and more negative findings made their way into journals. At the same time, questions were raised about the original study, and various methods that have been developed for identifying irregularities in bodies of research were applied to the money priming literature, indicating quite severe problems.[6] These methods are part of the set of tools, used in many scientific fields, called meta-analysis, which seeks ways of aggregating data from multiple studies. Intriguingly, an application in the field of parapsychology, the study of anomalous psychic phenomena such as telepathy, clairvoyance, and extrasensory perception, is widely recognized as the first modern meta-analysis. In the 1940s, the famous founder of parapsychology, J. B. Rhine, and his colleagues combined the results across over one hundred experiments on extrasensory perception, controversially concluding that it is a genuine phenomenon.

In combining multiple similar experiments to form an estimate of the average size of an effect, due heed needs to be paid to the possibility that the experiments that make their way into journal reports may not reflect all of the experiments that have been conducted. Suppose you conduct a test of extrasensory perception—for example, by asking a "sender" to look at a series of cards, each with one of four symbols on it, and a "receiver" to guess on the basis of the sender's transmitted thoughts which of the symbols is depicted on the card. Over a long series of trials and perhaps across many pairs of senders/receivers, you find that the receiver's accuracy is close to the level expected by chance, 25 percent. You write up your findings and send them to a prestigious journal such as *Science* or *Nature*. Your wait for a reply is likely to be brief and disappointing. If you think that academic journals exist for purely scholarly purposes, then reflect on the fact that Elsevier, one of the largest journal publishers in the world, made an annual profit of over $2 billion in 2021. Journals are a competitive business, and their publishers and editors strive relentlessly to increase their profile, readership, and revenue by publishing important and attention-grabbing scientific discoveries.

Your report with its low-key findings will not exactly get the editors excited. After trying with half a dozen other journals, you may decide to give up. In so doing, you have inadvertently illustrated the *file drawer problem*. This describes a bias in which the findings that make their way into the published scientific literature are incomplete, and not just incomplete in a random way: unsuccessful experiments and studies—the ones that fail to find a difference between two groups or some other meaningful difference—are much more likely to end up in the file drawer than successful ones. The inevitable consequence of this is that the published literature presents a biased and inaccurate glimpse of the truth. If only experiments that find extrasensory perception are published, while those (perhaps vastly more) that fail to find it are left hidden from view, we will end up believing something that's not true.

But if the failed experiments are languishing out of view in scientists' file drawers, how can we ever know that they exist? Counterintuitively, by examining studies that do get published, we see traces that are highly suggestive of the existence of unpublished ones. Rhine and his associates were among the first to devise methods for dealing with the file drawer problem, but since then, numerous more sophisticated tests have been developed, and when they are applied to money priming, they provide strong grounds for believing that around the world, many researchers' file drawers must contain failures to detect the expected priming effects.

The Telltale Signs of Publication Bias

Figure 9.1 illustrates one such test. Each black circle in the figure relates to one published money priming experiment.[7] The left-hand axis represents the precision or standard error of the experiment, a statistical concept that depends on how big the experiment's sample size is (the number of participants in the experiment). The axis is presented in reverse, such that studies with higher precision (because they used larger samples) appear toward the

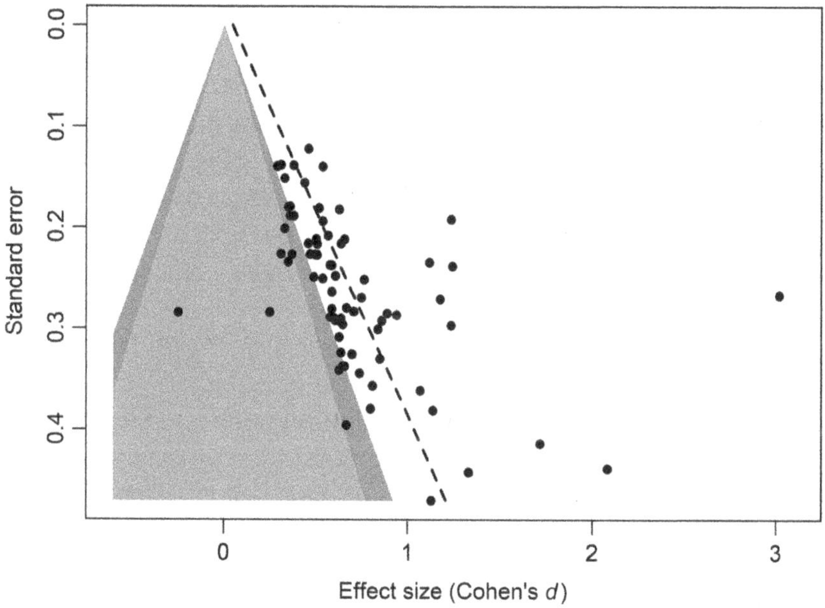

Figure 9.1

A funnel plot of data from a large number of money-priming experiments, each indicated by a dot. The vertical axis represents a measure that is related to the experiment's sample size (experiments higher up on the axis have larger samples). The horizontal axis represents the size of the money-priming effect (usually compared to a control group not shown the money prime) in a standardized measure, Cohen's *d*. Experiments finding a bigger priming effect fall further to the right. The figure clearly shows a relationship between these two measures: as an experiment's sample size gets smaller, its effect size increases. All experiments falling to the right of the gray funnel have statistically significant ($p < .05$) results, while those falling inside the funnel have nonsignificant ($p > .05$) results (the dark gray region indicates "marginally significant" effects falling between $p = .05$ and $p = .10$). The dotted trend line suggests that if an experiment is run with a very large sample, it will yield a priming effect close to 0 (apex of the funnel).

top of the graph. For example, the highest black circle on the graph comes from an experiment with a total sample of 275 participants, a reasonably large study, while the lowest point comes from an experiment with a mere 21 participants. The horizontal axis reflects the size of the effect obtained in each study. This is a common standardized measure known as Cohen's *d*, after the statistician Jacob Cohen, who pioneered many of our current approaches to scientific inference. For reference, think of a very obvious effect of one factor on a given measure, such as the effect of sex (male/female) on a person's height. The size of this difference measured by *d* is a little less than 2 in the global population. There is, of course, considerable variation in men's height and equally in women's height, but on average, men are taller than women. Cohen's *d* quantifies this difference relative to the variation among men and women. Most of the effects shown in the figure are rather smaller than this, as one would expect from behavioral research. An effect size of 0 reflects no difference between groups or no influence of a factor on the measure of interest.

It is clear that the majority of money-priming studies (in fact, all but one of the seventy-five experiments included in figure 9.1) yield a positive effect, meaning that they find that subtle suggestions of money make people work harder on boring or difficult tasks—or make them more selfish. If we were to average the effects sizes, it is clear that the resulting aggregate or "meta-analytic" effect size would be appreciably larger than 0. But this is not the key pattern in the figure; instead, it is the evident relationship between effect size (horizontal axis) and sample size (vertical axis). Studies with smaller samples tend to obtain larger effects, and the points tend to fall either toward the top left or lower right of this inverted funnel plot. This is not the pattern that we would expect. Larger studies should yield a more precise effect size estimate than smaller ones, but the points in the funnel should be distributed symmetrically. Think of this in terms of estimates of the average height difference between men and women. Very large studies should always yield estimates quite close to the true value (just under 2 in Cohen's *d* units). They may differ slightly due to random factors (the sample may by chance contain too many unusually tall women or too many unusually short men), but the large samples mean that such randomness should be averaged out. In contrast, small studies, including only one or two dozen individuals, for example, will inevitably yield noisy estimates of the true population difference, sometimes considerably overestimating and sometimes underestimating it, but the frequency of over- and underestimations should be about equal, giving rise to a symmetrical funnel shape.

What then explains the missing points in the figure, from experiments in which small samples yielded small effects (the lower left of the figure)? One answer is the file drawer effect: such studies exist but are languishing unpublished in researchers' filing cabinets. They are languishing there because they failed to yield a statistically significant effect, the famous *p*-value whereby a difference is only deemed to be "real" if (roughly speaking) the likelihood of obtaining a difference of that magnitude or greater by chance is lower than 1 in 20 ($p = .05$). The gray funnel area in the figure represents all combinations of effect size and sample size that yield statistically nonsignificant ($p > .05$) results (the dark gray region indicates marginally significant effects falling between $p = .05$ and $p = .10$). It is remarkable that the gray area so neatly separates a blank area where very few published findings exist from a cluster of published findings. In a nutshell, by examining published research, we can see a telltale pattern (asymmetry in the funnel plot) that is highly suggestive about the existence of unpublished research. This pattern is revealed only when we look at a large set of studies; it can't be seen in the individual studies themselves. This is a powerful demonstration of the importance of meta-analysis in research evaluation.

Of course in many situations we won't know anything in detail about these unpublished experiments, short of putting out a public call for scientists who work in the field to respond with information about any unpublished experiments they've undertaken on a given topic. Occasionally such calls are circulated, and indeed this has been done with respect to money priming.[8] What would we anticipate regarding these unpublished studies? Obviously the main expectation is that many of them were not published because they failed to obtain any sign of a money priming effect. (Others perhaps were unpublished for perfectly good but unforeseen reasons such as the researcher was unable to complete the experiment as planned or employed an outcome measure that proved to be unreliable.)

Is that what we see when we examine these unpublished experiments? Indeed it is. In another analysis of money-priming research, only about 35 percent of unpublished money-priming experiments obtain statistically significant results, compared to about 63 percent of published ones, and, moreover, unpublished ones yield an average effect size that is much smaller (about one-third the size) compared to published experiments.[9] A particularly striking confirmation of this relates to a form of priming that is a close cousin of money priming. In flag priming, a brief view of the American flag (it is claimed) unconsciously nudges individuals to be more right wing in their reported attitudes and voting intentions, even across a very long delay of eight months.

This quite eye-catching phenomenon was first described in a pair of experiments reported in 2011 by Travis Carter and his colleagues.[10] In a rather remarkable turn of events, Carter and his collaborators later (in 2020) opened up their file drawer to provide a frank peek into the publication habits of a single research team. In the years after their initial report, they conducted many more flag-priming experiments but published none of them. As they acknowledge, this was probably due to "motivated" reasoning—namely, finding reasons for not publishing the studies, reasons that happened to align with their motivation to believe that this form of priming is genuine. It is very easy for any researcher to tell themselves that an unsuccessful replication—perhaps conducted by a student new to the team and to experimental research—should be discounted because of poor execution or some other problem. Under this harshly revealing spotlight, the contents of Carter's file drawer make it clear how pernicious this bias can be: while their two published experiments obtained an average effect size of about $d=0.33$ (by the standards of behavioral science research, a meaningful effect), only one out of thirty-three unpublished experiments obtained a statistically significant effect and the average size of these effects was virtually zero. The fact that only the successful experiments were published means that the true status of flag priming was impossible to determine. When all experiments are combined, there is no priming effect in the totality of experiments conducted by Carter and colleagues, and other replications point to the same conclusion.[11]

This finding about money and flag priming is far from atypical. In a survey of over eighty meta-analyses in education and psychology that included both published studies and relevant unpublished ones solicited by extensive searches and well-broadcast appeals, the effect size calculated across the unpublished research was markedly smaller than that calculated across published research.[12] The conclusion is clear: if we focus only on research that makes its way past peer reviewers and editors and into journals and are not able to scrutinize unpublished research, we will be looking at a biased and unrepresentative snapshot of the truth: studies that have been cherry-picked on the basis of obtaining positive effects. The peer-review process has many virtues and helps to weed out poor-quality research, but it also introduces an unintended bias: published research will often present an overly optimistic picture of the evidence.

Indeed the cherry-picking in the case of money priming is so extreme that it miraculously turns a noneffect into an effect. Like flag priming, there is almost certainly no money-priming effect in the conditions that prevail in these experiments. If we fit a trend line to the data points in figure 9.1, we can ask what the expected money-priming effect would be in an experiment with a very large sample, that is, one with a standard error near 0. The dotted line in the figure clearly suggests that this effect size would be very close to 0—the line almost touches the apex of the gray triangle, indicating a Cohen's *d* of 0. This seems like a bit of magic: from a set of experiments, almost all of which find a positive money-priming effect, we can extrapolate to what the effect would be in an "ideal" experiment, and determine that this effect would be negligible. In the next chapter, we will see that preregistered experiments designed from the outset to eliminate any possibility of bias confirm that money priming is not a genuine effect.

Money and flag priming are just two examples of a class of effects that have been labeled "social" or "behavior" priming. Other varieties, also with catchy names, include "intelligence" priming (in which individuals answer more general knowledge questions correctly when they previously thought about what it would be like to be a professor), "romantic" priming (images or text about romantic situations make men more willing to take risks), "religious" priming (subtle activation of the concept of God renders people more willing to behave prosocially), and many others (you get the idea). These represent controlled laboratory experiments that model everyday situations in which subtle cues or events might nudge our behavior unconsciously, like the claim that in-store aromas motivate us to spend more money. Like money priming, these other effects have not withstood closer scrutiny and are probably nonexistent.[13] The purported demonstrations of these effects likely represent spurious findings contaminated by publication bias and the creative employment of researcher degrees of freedom.

Researcher Degrees of Freedom

The existence of unpublished experiments obtaining smaller effect sizes than published studies is not the only possible explanation for the asymmetry seen in the funnel plot. Another possibility is that researchers might engage, perhaps inadvertently, in practices that exploit so-called researcher degrees of freedom, another avenue for bias to enter the research process.[14] Consider the following seemingly innocuous scenario. A researcher is interested in whether there is a difference in behavior between two groups, perhaps in a money-priming experiment. For one group, subtle reminders of money are shown, whereas for the control group, they are not. The researcher measures (perhaps via a questionnaire) how willing participants assigned to the two groups are to engage in some volunteering activity. Suppose that there is no true priming effect. Although most of the time the experiment will correctly find no difference between the groups, occasionally—purely as a result of random fluctuations in the data—it will spuriously find a priming effect. Initially the researcher recruits twenty participants for each group and then examines the data, finding that her hypothesis seems to be confirmed: willingness to volunteer is lower in the group primed with reminders of money. However, this effect is quite small and doesn't reach the conventional threshold for statistical significance. The researcher is confident that her observed effect is genuine and that topping up her groups will reach the statistical significance threshold. She therefore tests another twenty participants in each group, reanalyzes the resulting data (now with forty participants per group), and finds a difference in willingness-to-volunteer scores that now meets the $p < .05$ threshold. She writes up her results for a prestigious journal.

The problem is that, innocently, the researcher has exploited a researcher degree of freedom (in this case, deciding how many participants to test in each group, the sample size, based on the results) in such a way as to bias her findings. Suppose the difference hadn't reached statistical significance after forty participants per group; she would probably have tested yet more, and so on until exhausting either her pool of participants or her patience. But clearly this "optional stopping" procedure inflates the probability that a purely random difference between the groups will emerge and be mistaken for a true difference. Indeed if carried on indefinitely, this procedure of repeatedly topping up and peeking at the data is guaranteed to yield a statistically significant difference, even when none exists in the population, because the experimenter will inevitably encounter one of the random fluctuations and end up capitalizing on chance rather than detecting a genuine effect. If you throw a dart once at a small target, the chances of hitting it are low. But if you throw the dart one hundred times, you would be very unlucky not to hit the target eventually.

Or consider another way in which flexibility in carrying out an experiment can lead to spurious findings. Imagine that another researcher measures how hard participants are willing to work on a boring task. After testing twenty participants in the money-primed and control groups, he observes a small but statistically nonsignificant difference in the number of minutes participants in each group are willing to work on average. Noting that the difference is in the expected direction, he looks carefully at the individual data and sees that one participant in the control group works for an unusually long time while one in the primed group works for an unusually short time (the hypothesis being that money primes people to work harder). He therefore treats these data points as outliers (following perfectly sound statistical practice for excluding rogue data) and drops them from his analysis. Now the group difference reaches the magical $p < .05$ threshold, and he writes up his results for a prestigious journal.

This researcher has also exploited a researcher degree of freedom—in this case, the precise rule for treating observations as outliers. With many different choices that can be made regarding the precise outlier rule, as well as other similar decisions about transforming the data (again good statistical practice), he is boosting the probability that a random difference will look like a meaningful, nonrandom one. Some rather evocative terms are often used to refer to the many choices researchers can make that can increase the likelihood of obtaining an apparent effect in their data, even if no such effect exists, as a result of decisions taken after observing the results. One is "*p*-hacking," meaning the various tricks that a researcher can try to push a set of data over the magic $p < .05$ threshold. Another is the "garden of forking paths," from the title of a story by the Argentine writer Jorge Luis Borges, by which statistician Andrew Gelman characterizes the numerous different pathways researchers can take in analyzing their data, some of which might lead to spurious differences being obtained.

Whatever one calls these practices, they have the consequence of moving an experimental result that "should" be inside the gray funnel in figure 9.1 to a location outside the funnel.

While researcher degrees of freedom and *p*-hacking are descriptive of particular behaviors on the part of scientists, well-known concepts such as confirmation bias and motivated reasoning may be invoked to explain psychologically why these behaviors occur. When researchers set aside failed experiments and consign them to their file drawer but publish their successful experiments, they may be falling prey to confirmation bias—the tendency to search for and favor information that supports one's beliefs in preference to disconfirming information. One particular variety of this bias takes the form of experimenter expectancy effects. The experimenter believes so strongly that a particular outcome will occur (and perhaps even wants that outcome to occur) that they unintentionally influence the participants in the study to conform to that expectation. You may recall the discussion in chapter 3 of research showing that experimenters who expected research participants to walk more slowly down a corridor indeed observed this outcome, and it is precisely to avoid such effects that double-blind procedures—in which both experimenters and participants are kept unaware of information that could bias their behavior—are employed in most medical trials. Expectancy effects are rife in behavioral research, including in experiments on priming.[15]

As several surveys have documented, many researchers (ourselves included) admit to having carried out practices at some stage in their careers that exploit researcher degrees of freedom. We emphasize that these practices can be and usually are entirely innocent; a researcher can in all honesty believe (and have good grounds for believing) that increasing the sample size will allow a true effect to emerge or that an observation is an outlier. Indeed our academic mentors sometimes positively encourage us to do so. We encountered Daryl Bem in the previous chapter in the context of his evidence that participants in his experiments could predict what picture was about to be displayed on a computer screen. Aside from this controversial work, Bem is a social psychologist famous for many ground-breaking contributions to research on topics like cognitive dissonance. He wrote the following in an influential guide to student researchers on how to write a journal article:[16]

> Examine them [the data] from every angle. Analyze the sexes separately. Make up new composite indexes. If a datum suggests a new hypothesis, try to find additional evidence for it elsewhere in the data. If you see dim traces of interesting patterns, try to reorganize the data to bring them into bolder relief. If there are participants you don't like, or trials, observers, or interviewers who gave you anomalous results, drop them (temporarily). Go on a fishing expedition for something—anything—interesting.
>
> No, this is not immoral. . . . In the confining context of an empirical study, there is only one strategy for discovery: exploring the data. Yes, there is a danger. Spurious findings can emerge by chance, and we need to be cautious about anything we discover in this way. In limited cases, there are statistical techniques that correct for this danger. But there are no statistical correctives for overlooking an important discovery because we were insufficiently attentive to the data. Let us err on the side of discovery.

This sounds strikingly like a call to undertake a fishing expedition with one's data in every conceivable way until an interesting pattern emerges. This is exactly the kind of behavior that can introduce bias into the research process and increase the likelihood of spurious findings. To be fair to Bem, he does make it clear in his guide that he is referring to *exploratory*, discovery research where novel hypotheses are being formulated and new insights sought, and he emphasizes that this is different from *confirmatory* or justificatory research, where a clear hypothesis is being put to the test and all data analysis decisions are made in advance of seeing the results, thus reducing the chances of bias. It is indeed perfectly reasonable to probe one's data in every conceivable way in the search for a brilliant new discovery or insight, provided one is transparent about doing so and the crucial pattern is replicated and confirmed in a purely confirmatory follow-up study. But he goes on to recommend that "the data may be strong enough to justify recentering your article around the new findings and subordinating or even ignoring your original hypotheses." Nothing could illustrate the crisis of scientific credibility better than this advice to present exploratory research as if it's confirmatory.[17]

What Bem is recommending is the practice known as HARKing, for hypothesizing after the results are known.[18] HARKing means reporting a hypothesis that in reality emerges from a set of data as if it were formulated before the data were collected. It is bad science because it can radically change the credibility of a pattern in a set of data. If I hypothesize in advance that a money prime will render people less willing to help others and my experiment confirms this prediction, then the hypothesis rightly gains considerable support. It would then be wholly reasonable to expend time and effort weaving the hypothesis into a larger theoretical framework. But if the hypothesis was only derived after the data were analyzed—and perhaps the data had to be massaged in complex ways before emerging—then it gains almost no support from the data. The data cannot both form the basis of the hypothesis and provide support for it. This would be circular.

In a nutshell, there are many ways in which scientists can run their experiments and analyze their data, and the ensuing garden of forking paths means that they are highly likely eventually to find something spurious in the data that looks meaningful and (more important, from the scientist's point of view) publishable, even if in reality what they've "found" doesn't exist. We contend that this is what has happened in many areas of research on the unconscious, but we emphasize that these problems probably exist across the entire breadth of the sciences. There is abundant evidence from surveys that chemists, biologists, medical researchers, and those from many other disciplines admit *p*-hacking. The net result is that a high proportion of "findings" in science are likely to be misleading or even outright false.[19] Money priming provides a compelling example.

Although the evidence is striking, it is somewhat indirect. The asymmetry of the money-priming funnel plot strongly points to publication bias and the exploitation of researcher degrees of freedom, and researchers' responses to surveys make it fairly clear that questionable research practices are rife, but nonetheless these forms of evidence fall short of demonstrating concrete, irrefutable examples of poor practices such as *p*-hacking.[20] Fortunately we don't have to rely solely on these arguments, as there are now unequivocal illustrations of *p*-hacking in many specific pieces of research. One group of investigators took advantage of the fact that platforms for distributing questionnaires and collecting survey responses sometimes require all questionnaires and data to be made publicly available.[21] Hence it is possible to compare the eventual published journal article reporting a survey against the complete questionnaire that was administered. This contrast yields a stark outcome. A sizable proportion of published studies failed to mention all of the different experimental conditions in the survey. Why would this occur? The obvious reason is that a condition failed to yield the findings that the researchers expected, and they conveniently omitted it from their report. Even more startling was the finding that a majority of studies failed to report all of the measures collected in the survey, again presumably because the results were inconveniently at variance with the researchers' expectations and didn't fit into the nice story they wanted to tell. If experiment 1 yields several results that fit with the researcher's theory, but experiment 2 confirms only some of these results, then how convenient is it to pretend that the unwelcome negative findings in experiment 2 just didn't exist and that the experiment never tested these outcomes?

In addition, the results that did make their way into journal publications were much more likely to reach the $p < .05$ threshold for statistical significance than those that did not. This is rather incontrovertible evidence that researchers cherry-pick the findings that fit into the story they want to sell. When results fail to support their hypotheses, they disappear as if they were never part of the study.

Analyses in the Multiverse

In the standard scientific publishing model, researchers carefully describe their methods and then go on to explain the findings and their statistical interpretation of those findings, but only a single analytic method is described. The researcher chooses a single rule for dropping participants from the study, chooses a single way of dealing with outlier observations, chooses a single statistical test, and so on. Each of these choices offers scope for the exploitation of researcher degrees of freedom or *p*-hacking. Another growing trend to minimize the harm of *p*-hacking is to report the effects of making different decisions at each of these choice points, in what is called a "multiverse"(or "sensitivity") analysis. If a finding is genuine and robust, then it should still be evident even when all sorts of different choices are made about how to analyze the data. Conversely, a finding that depends critically on one specific set of choices (one route in the garden of forking paths) and disappears if any of these choices is changed is not a robust one that should be relied on for theory or practice.

Consider the following simple question: Are referees in soccer matches unconsciously more likely to give red cards to darker-skinned than to lighter-skinned players? A red card results in the ejection of the player from the game as a punishment for a major rule violation or unacceptable aggression. It has long been suspected that racial biases play a role in such decisions, but how might one go about testing this claim? Raphael Silberzahn from the University of Sussex Business School and his colleagues set out to answer this question in an unusual way by relying on a multiverse analysis.[22] First they created a data set based on information from a sports statistics company. In this data set, information on over fifteen hundred top-division players included their skin tone (judged from photos) and their interactions over the course of their careers with each of over three thousand referees (in particular, red cards given), as well as a range of details about each player's age, height, weight, and so on. One might think that on the basis of this data set, it would be fairly straightforward to determine whether darker-skinned players received more red cards than lighter-skinned ones.

But a moment's reflection suggests that there will be quite a few forking paths in this particular garden. For example, it might be the case that darker-skinned players tend to play slightly more often in defensive positions than lighter-skinned ones (or vice versa). Defenders might be slightly more (or less) likely than attackers to be given red cards. These two trends, which might be very slight, could nonetheless combine to yield a pattern in which darker-skinned players falsely appear to be more prone to punishments, but the pattern would not be indicative of an influence of skin color. So any approach to analyzing this data set will inevitably throw up a range of questions that the investigator needs to address. In an ingenious approach, Silberzahn and his colleagues simply invited expert research teams from around the globe to take on the challenge of analyzing the data set according to their own particular preferred approaches, and twenty-nine agreed to take up the challenge.

What was the outcome? No two teams reached exactly the same estimate of the effect of skin tone on the likelihood of being given a red card, and although the majority (about two-thirds) of the teams concluded that there is a relationship, many (about one-third) concluded that there isn't. The teams adopted a staggering range of analytic approaches, using numerous different statistical techniques.

The salutary point of this example is that any one of the analyses could have been undertaken individually and justifiably published in the normal way in a peer-reviewed journal. A total of twenty-nine articles would have made their way into the literature, with no clear consensus about the true answer to the question. Since all the teams analyzed exactly the same data set, we can say categorically that variation in the decisions that intelligent researchers make about how to analyze their data can lead to polar opposite conclusions. If we have no transparency about these decisions and about how robust researchers' conclusions are in the face of different sets of decisions, then we cannot reasonably evaluate the outcome of any single piece of research.[23]

The convergence of the biases discussed in this chapter yields spurious conclusions about the mind. In case the evidence we've presented isn't sufficient to convince you of this, then consider one final point. Against any reasonable scientific criteria, the quantity of evidence for paranormal phenomena such as telepathy, clairvoyance, and precognition (seeing the future) is overwhelming. Hundreds of research reports have been published in peer-reviewed journals of successful demonstrations of these phenomena. In the previous chapter, we described Daryl Bem's infamous experiments apparently showing that people can know, in advance, where an erotic image was about to be displayed on a computer screen. Although we described many reasons (quite apart from their implausibility) not to believe Bem's findings, the general claim that paranormal phenomena exist rests on vastly more evidence than this one set of dubious experiments. It seems highly likely that if we collected a large amount of data relating to some putative but in reality nonexistent paranormal phenomenon and subjected those data to a multiverse analysis, at least some of the analysts would wrongly conclude that the effect is genuine.

Etzel Cardeña, a psychologist at Lund University in Sweden, has summarized meta-analyses and concluded that they provide compelling support for telepathy, clairvoyance, precognition, and related phenomena.[24] Indeed the evidence from these meta-analyses is probably—by any objective standards—at least as strong as the evidence for many standard psychotherapy treatments and numerous other widely accepted results. Cardeña takes them as proving the existence of the paranormal. For anyone with a more skeptical view of such phenomena, they confirm the existence of a raft of research practices and biases that allow scientists to fool themselves and others.

10 Research Reformed

It is plain that selective reporting, *p*-hacking, and other common practices can lead to gross distortion of the scientific record, with published results commonly presenting a false picture of reality. An obvious question at this point is why the practice of replication (repeating a previous study's methods to obtain new data and see if similar results are obtained) does not rapidly weed out spurious findings. As we noted in the previous chapter, anyone trying to market a nonfunctional new type of battery would instantly be found out. Incorrect scientific or technological developments do not last long under the glare of immediate feedback. When chemists Martin Fleischmann and Stanley Pons announced in 1989 that they had produced cold fusion, the prospect of almost limitless clean energy galvanized the public and media. But within a few weeks, after many independent teams had failed to confirm Fleischmann and Pons's findings, the *New York Times* declared that cold fusion was dead. If money priming is not a real phenomenon, then why didn't failed replications immediately reveal this?

Replication and Registration

At various points in this book, we have discussed examples of replication failures. In chapter 3, we described two famous psychological experiments (walking slowly and smiling through your teeth), neither of which proved to be replicable. In chapter 4, we briefly reviewed another one, on incidental anchoring (people don't pay more at restaurants with high numbers in their names). Because the results of individual studies can be incorrect due to flaws or random error, replication is fundamental to confirming the validity of a scientific claim. Indeed we could go further and imagine a world in which every published result in science was ignored by other scientists, as well as by the media, until a successful replication was reported. The truly terrible consequences of the false link between the measles, mumps, and rubella vaccine and autism would never have happened, for example.[1]

Unfortunately, direct replication plays a tiny role in most scientific fields. Estimates consistently suggest that only around 1 percent of all psychological research is ever replicated, a state of affairs that is almost universally recognized as needing to change.[2] Part of the problem is that, as we discuss later, science is a competitive field, and researchers often think that they will receive little reward for investing time and effort into "merely" reproducing a result that is already known. After all, the glory goes to the person who first made the discovery, not the unimaginative drudge whose contribution is simply and boringly to confirm it. But in light of what we now know about publication bias and *p*-hacking, researchers are starting to undertake more and more replications, particularly of eye-catching results.

Money priming is one such result, and a major replication study led by Richard Klein of the University of Florida sought to reproduce it in a very large-scale, multilab project.[3] Thirty-six teams from around the world agreed to participate, each running the same battery of tests designed to generate thirteen well-known effects, including money priming. With a near-identical procedure to one of the original experiments demonstrating the effect, participants began by answering some demographic questions via computer. For some of them, the background was a faint picture of $100 bills, while for others, it was a blurred version of the picture in which the bills could not be identified as such. Then participants answered questions regarding their attitude to the fairness and legitimacy of the prevailing social system. A typical item was, "Everyone has a fair shot at wealth and happiness," rated from "strongly disagree" to "strongly agree."

Figure 10.1 reproduces the funnel plot from the previous chapter, but now adds the thirty-six individual results from this multilab replication project, as well as those from another replication effort, indicated by open triangles.[4] A couple of things are immediately apparent. First, these replication results are generally higher in the figure than the original studies. This means that they yield more precise estimates (they have smaller standard errors), which in turn comes from the fact that they employed substantially larger samples than the original experiments: while the original experiments tested a median sample of only 66 participants, the replications had a median

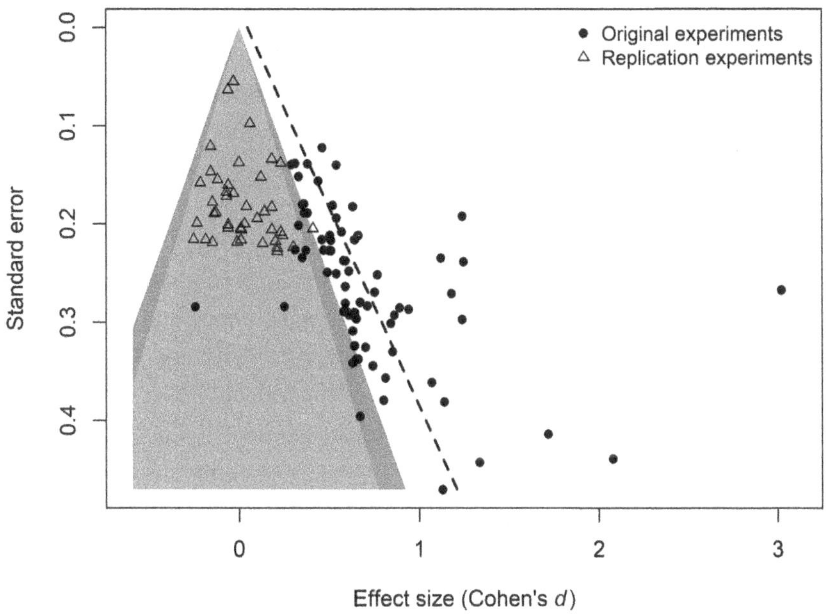

Figure 10.1
This figure reproduces figure 9.1 but adds the results of preregistered replication experiments (open triangles). The vertical axis represents each experiment's precision, while on the horizontal axis is depicted the outcome of each experiment, measuring the size of the money priming effect in the standardized measure, Cohen's *d*. Experiments falling to the right of the gray funnel have statistically significant ($p<.05$) results, while those falling inside the funnel have nonsignificant ($p>.05$) results (the dark gray region indicates marginally significant effects falling between $p=.05$ and $p=.10$). The dotted trend line is fitted to the original experiments only.

sample of 110. This is still not very large by the standards, say, of medical trials, but at least it is a step in the right direction. The second obvious aspect of the replication results is that they fall symmetrically within the funnel. Unlike the asymmetry of the distribution of original effects, the replications show no tendency for a relationship between precision and effect size. Finally, and most important, they cluster around an effect size of 0—that is, no overall difference in the attitudes to the prevailing social system of participants primed with money compared to those not primed. Indeed there is almost no overlap in the effect sizes of the original and replication experiments. Despite there being literally hundreds of studies appearing to obtain money priming effects, a near-exact replication project failed completely to detect an effect.

How can this be? The replications vindicate the conclusions of the funnel plot discussion in the previous chapter and bolster the inference that many of the apparently successful demonstrations of money priming are false positives—results appearing to find an effect that is not real in the population—resulting from *p*-hacking or good fortune (the researchers ran many studies, and the published ones are those that by chance found a statistically significant effect). The many "missing" studies in the funnel plot are the telltale clue attesting to this. But one might ask why the replication studies are any more believable than the original ones. Don't we simply have a case here in which one set of studies disagrees with another set? After all, the fact that Klein's replication project was conducted after many of the original studies were reported is neither here nor there: the original studies fail to replicate the null findings of the replications to just the same extent that the replications fail to replicate the positive findings of the original experiments.

To see how implausible this interpretation is, remember that thirty-six independent teams contributed to the replication, and these teams had a diverse set of expectations about what they would find, some believing the effect would be found and others not. If special expertise and care or expectations are needed to obtain the effect, then at least a few of the teams should have found a positive money-priming effect, but in fact only one team did—just as would be expected by chance bearing in mind that $p < .05$ implies a lucky positive result once in every twenty or so attempts.

But the strongest reason to believe the replication findings over the original ones is that Klein and his colleagues preregistered their entire study. Before collecting any data, they carefully described exactly how their study would proceed and how the data would be handled and analyzed, effectively tying their own hands to prevent any possibility of later *p*-hacking. As promised in the preregistration, the data were not examined until all testing had been completed. The preregistration was uploaded to a public repository together with the program for the experiment itself in advance, so anyone can go back and check that they did exactly what they said they would do. These features are in stark contrast to standard experimental practice. For each of the "biased" studies in the funnel plot (the original experiments), we simply do not know whether multiple analyses were run and only the significant ones published; whether participants were added or removed after running initial, exploratory analyses; or whether these studies are only a subset of all the studies ever conducted by those teams (and we can't know what those other unpublished studies would look like). In preregistered studies, in contrast, what you see is all there is.

Preregistration is rapidly becoming a crucial method for boosting the credibility of research, going a long way to eliminating many of the evils discussed previously.[5] While *p*-hacking is the most obvious one, others are eliminated as well. Because the preregistration describes the experimental hypothesis in detail, the researcher's ability to indulge in flexible retrospective HARKing (reinterpreting a surprising result as if it were predicted all along) is severely curtailed. Publication bias is also appreciably less likely, not only because the preregistration is published in the sense of being a publicly accessible document, but also because as long as the study was executed in accordance with the stated plan, its findings are likely to be a contribution to the academic literature: if it replicates the finding it was attempting to repeat, then it's a valuable confirmation of that finding, whereas if it fails to replicate the earlier result, that itself is important knowledge.

Preregistration powerfully emphasizes the crucial distinction between two forms of research endeavor briefly mentioned previously: exploration versus confirmation. Exploratory research is what we all have in mind when we think of a scientist working at a laboratory bench, trying to make a discovery, solve a problem, or build a new device. Exploration is unquestionably the engine of scientific and technological advancement, as well as being the main yardstick against which scientists themselves are judged. But confirmatory work—carefully seeking to validate previous claims and findings—is just as important. It is an essential tool for us to separate out true findings from all the *p*-hacked false ones. Indeed some have argued that psychological research in general should move toward a model in which research publications comprise initial exploratory studies followed by large-scale confirmatory ones.[6] But what does an ideal confirmatory study look like?

In medical research, it has been compulsory for many years to preregister clinical trials before conducting them. For example, ClinicalTrials.gov, established in 2000, is a repository of, to date, about 400,000 trials. Laws mandating the registration of trials involving drugs or devices have been passed in both the United States and the European Union. This sounds like an excellent mechanism to decrease the chance that the public will be exposed to treatments or drugs that are in fact ineffective. Surely the researchers conducting the trial cannot *p*-hack the results in order to gain a statistically significant result if their hands are tied by their preregistered commitments about how they would conduct and analyze the trial? And indeed there is evidence that clinical trials are becoming less and less successful.[7] This may sound like bad news, but in an important sense, it's the exact opposite. We have argued that much of the published scientific literature comprises false-positive results, either wrestled out of unpromising data by expert *p*-hackers or simply the lucky survivors of the Darwinian selection process that diverts successful studies into scholarly journals and unsuccessful ones into the file drawer. If this is even remotely correct, then we should expect any mechanism that suppresses *p*-hacking and publication bias to decrease the number of false positives in the scientific record. So the fact that fewer and fewer trials in some domains are succeeding may, paradoxically, be a good sign.

Unfortunately this form of preregistration does not provide any ironclad guarantee that research practices will improve. It fails to protect against publication bias because the researcher may choose not to submit or a journal may choose not to publish the results if they are messy or negative. Moreover, and somewhat amazingly, analyses show that researchers engage in widespread *p*-hacking even when their public preregistered methods descriptions make it easy for anyone to spot the *p*-hacking. A prime example of this is the switching of a trial's designated primary outcome. Imagine that a trial is being run to measure the efficacy of a new medicine in treating headaches. As part of the preregistration, the researcher may announce that the number of headaches per week is the crucial outcome measure, the one on which the trial's success stands or falls. This measure either shows a statistically significant decline, in which case the trial has been successful, or it doesn't. Later, the trial is published in a scholarly journal, but now the key outcome measure that the article analyzes is headache duration, not number. The researcher has switched outcomes between preregistration and publication of the results. Obviously a likely explanation is *p*-hacking: the effect was statistically significant on the duration but not the number measure, so the researcher switched them in order to get the article published. The scale of these reporting switches is alarming, and consistent with the *p*-hacking explanation: when outcomes are switched, they overwhelmingly tend to be in favor of achieving statistical significance.[8]

These switches also serve to highlight (if we needed further evidence of this) that the peer review process falls far short of providing a guarantee that published research is credible and maintains high standards of research probity and rigor. One might hope that reviewers would immediately spot these switches and other deviations from the preregistration and reject the paper for publication, but this rarely happens. Peer reviewers are unrewarded for their work and have little incentive to spend undue amounts of time cross-checking a manuscript against a preregistration, and indeed there is concrete evidence that they rarely do so.[9] Preregistration is a step in the right direction and at least means that *p*-hacking becomes visible to anyone wishing to compare a preregistration against published results, but it is not a cast-iron method of boosting research credibility.[10]

A stronger form of preregistration is beginning to gain a foothold and may prove to be the ideal format for conducting confirmatory research. In so-called *registered reports*, the researcher describes in complete detail how she plans to carry out a study or experiment, as well as the hypothesis being tested, the primary outcome measure, the data analysis method, and so on.[11] But instead of simply posting this description on a time-stamped public repository and then proceeding to collect the data, as would be the case under standard preregistration, she instead submits the description to a journal for evaluation. The journal asks reviewers to assess the described study for its rigor (for instance: Will its sample size be adequate? Is the method appropriate to test the hypothesis?) and likely contribution to the field, and then if it is judged of sufficient quality guarantees to publish it once the study has been completed, *regardless of the results*. The journal is in effect making a results-blind decision about the work that places all the emphasis on the rationale and methodological rigor of the study and none on the results. The results will be what they will be. In the process of approving the final article for publication, the journal reviewers are asked to check that the researcher conducted the study according to the preregistration description, has explicitly noted any deviations, and clearly flags any new analyses that were not preplanned as exploratory ones not to be confused with the primary confirmatory analyses.

It's easy to see that registered reports of this form, which are now solicited and published in many journals, provide a high level of protection against selective publication, *p*-hacking, HARKing, and so on. The results are published regardless of what they reveal, hence protecting against the selective nonreporting of statistically nonsignificant results. The researcher precommits to the analysis, eliminating the scope for *p*-hacking. And because the hypothesis is stated in advance, there is minimal scope for seeing the results and then going back to change the purpose of the study.

Indeed it is now becoming apparent that the quality of research published in registered reports is appreciably higher than in standard peer-reviewed journal articles. In a recent project led by Courtney Soderberg from the Center for Open Science, a nonprofit organization based in Charlottesville, Virginia, over three hundred experts were recruited as assessors.[12] Each was given a deidentified and lightly redacted registered report (thus reducing the likelihood that the assessor realized that it was a registered report) as well as a carefully selected and matched standard journal article to evaluate. Across nineteen evaluation criteria, the registered reports scored higher on all dimensions. Their methods and analyses were rated as more rigorous, they were judged more novel and creative, the quality of the discussion of the findings was judged better, and so on. As a means of enhancing research quality (as well as boosting public trust in science), it would not be an exaggeration to suggest that future generations will look back at the registered report format as one of the most significant methodological developments in the history of science.

There is a clear further test of the idea that registered reports provide protection against publication bias and *p*-hacking: they should yield positive findings much less frequently than standard nonregistered publications. Put the other way around, null results should be much more common in registered reports than elsewhere. This issue, which we've already touched on in relation to unpublished studies and clinical trials, is a key indicator of the credibility of the scientific literature. We know that the vast majority of published studies report positive—in other words, statistically significant—findings. Across social and behavioral research, estimates of the proportion of null results vary but are generally in the range of 5 to 20 percent.[13] What do we find when we look at the rate of null findings in properly preregistered experiments? It is startlingly higher. Although the researchers carrying out these registered experiments have framed plausible hypotheses and tested large samples of participants, their carefully and publicly preplanned experiments are successful only in a minority of cases in finding a meaningful effect or group difference. In at least half of all cases, they yield a null result.[14] These findings tell us loud and clear that the high rate of positive findings in the normal scientific literature (at least 80 percent) cannot be a true and unbiased reflection of the world.

The money-priming literature provides a perfect illustration of this difference. In a comprehensive meta-analysis of all available studies,[15] including 47 preregistered tests and 189 standard non-preregistered ones, 62 percent of all standard studies obtained positive results but only 11 percent of the preregistered ones did. This same pattern can be seen in the funnel plot shown in figure 10.1. The open triangles all come from preregistered studies and find effects close to 0, whereas the other points reflecting standard experiments are much more likely to indicate positive effects. Recall that the gray area in the funnel depicts all effects that are statistically nonsignificant.

There is one further consequence of the greater prevalence of null results in preregistered experiments, combined with the growing frequency of such experiments: we should see many effect sizes dwindling over time. If early reports are biased by *p*-hacking and publication bias, whereas later preregistered studies ameliorate these biases, then observed effects should become smaller and smaller, and this will be true whether they are genuine effects or completely spurious. In the former case, the effect size will eventually converge on the true positive estimate. This "decline effect" is what has happened with studies on cognitive-behavior therapy (CBT) for depression, for instance, with the observed efficacy of this type of treatment slowly dwindling over the past forty or so years.[16] Despite this, the most up-to-date estimates still show it to be quite effective.

In the case of truly spurious effects, we would expect the estimated effect size to eventually converge on 0. The final nail in the coffin of money priming is that studies show exactly this pattern. Figure 10.2 graphs the effect sizes of money-priming tests, including both published and unpublished studies, some preregistered and some not, across time. Although the data go up to only 2017, it is clear that the effect has been declining steadily since the original studies in 2005 and 2006. The best estimate of the outcome of a money-priming study conducted after 2018 is very close to zero. After all this huge effort studying an eye-catching way of unconsciously nudging people's behavior and the vast amount of journal space devoted to it (not to mention the many taxpayer-funded research grants), we find that the effect proves to be no more real than the telepathy, clairvoyance, and extrasensory perception effects first studied experimentally in the 1940s by J. B. Rhine.

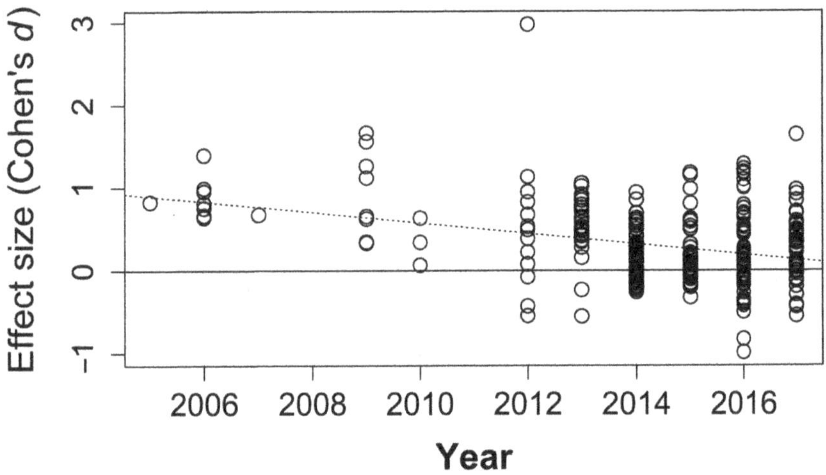

Figure 10.2
This figure depicts, across each year since the original money-priming experiments were published, the effect size (in Cohen's *d* units) of every test that has been conducted (the data were compiled by Paul Lodder). Positive effect sizes represent effects in the direction expected by the money-priming hypothesis (for instance, that subtle reminders of money cause people to work harder on difficult tasks). Remarkably, up until 2012, every study yielded an effect in the predicted direction. Since then more and more replications yielding null results, including preregistered studies, have been conducted, including those from the multilab project led by Richard Klein that contribute thirty-six of the data points scattered close to 0 in 2014. The dotted trend line shows that the overall effect has been steadily declining since it was first reported.

The Power of Myths

There is another reason that eye-catching claims in psychology, including ones like money priming that concern unconscious mental processes, often maintain high prominence and uncritical acceptance long after they have been discredited. It is because we are strongly persuaded by story lines and myths that make sense of the complex and confusing world around us. Saying that many aspects of our behavior arise through unconscious influences seems a simple and parsimonious way of explaining actions that would be fiendishly hard to rationalize any other way. If you were asked to explain why you worked unusually hard on a given task—cleaning the house one day, say—you would probably struggle to come up with a convincing explanation in terms of the conscious thoughts and motivations that went through your mind at the time. But saying that you were influenced by numerous factors of which you were largely unaware—such as your eyes fleetingly glancing at a banknote lying on the table—provides a compelling story. You don't even have to enumerate all the factors that influenced you. That's part of the beauty of the explanation—that these factors were unconscious and hence you can't report them.[17]

And this storytelling is part of science itself. Take the case of citations to research that has been strongly contradicted in replication efforts. When scientists write up their research for journal publication, they introduce their project and its purpose by reference to previous relevant work. This introduction section, together with the discussion presented at the end of the article once the results have been described, tries to present a narrative that makes sense of the results and fits them into a larger perspective on the topic. In citing relevant previous research, one would expect a fair-minded approach in which the strengths and weaknesses of the earlier work are evaluated from a neutral perspective, regardless of whether the cited work fits in with or runs counter to the author's own perspective. When we look at citation patterns, we soon see that this assumption is a gross idealization. Particularly striking is that high-profile original studies continue to be cited at high rates even if they've been strongly contradicted by subsequent replications. Money priming provides a clear illustration.[18] Despite the fact that Klein's large-scale, multilab project completely failed to replicate money priming, researchers continue to cite the original research by Vohs and her colleagues almost as if the replications didn't exist. In the five years following the publication of Klein's failed replication, the number of annual citations of one of the key original reports continued unabated. One might imagine that later researchers were citing the original report in the context of discussion about its unreplicability, but this was not the case. The vast majority of these citations were favorable, discussed money priming as if there was no problem with it, and did not cite the Klein article. Moreover, across several case studies of this type, even in those instances where authors did cite both the original research and the replication failure, they often provided no explicit justification for their favorable assessment of the original research.

Scientific textbooks, one of the main ways in which knowledge in a discipline is transmitted to new members of the discipline (that is, students) and interested laypeople, provides another illustration of how myths can lead to distorted evaluation of research. Numerous case studies document the myths that textbooks help to sustain. A fascinating and instructive one concerns what is probably the most famous experiment in all of psychology, the Stanford prison experiment. In August 1971 Philip Zimbardo assigned students by a coin toss to the role of "guards" or "prisoners" in a mocked-up prison in the basement of the Stanford University Psychology Department, with Zimbardo playing the role of prison superintendent. The experiment had to be shut down after six days because the students adopted their roles rather too convincingly. The guards started to commit acts of psychological torture on the prisoners, some of whom accepted their roles as victims of abuse. The experiment is widely taken as providing evidence that the context (including the social roles placed on us) plays a far greater role in determining human behavior than individual personal dispositions such as our particular personality attributes.

But this is little more than a story.[19] From a scientific point of view, the Stanford prison experiment comes nowhere close to demonstrating the power of social roles. The guards didn't act as they did because of their roles as guards, but because Zimbardo effectively instructed and guided them in how he expected them to behave. Subsequent reports from those who took part make this abundantly clear. Carlo Prescott, an ex-convict who served as chief consultant to Zimbardo on real prisons, later said that "Zimbardo began with a preformed blockbuster conclusion and designed an experiment to 'prove' that conclusion." John Mark, one of the guards in the experiment, commented that Zimbardo "knew what he wanted and then tried to shape the experiment. . . . He wanted to be able to say that college students, people from middle-class backgrounds—people will turn on each other just because they're given a role and given power." Most striking, in later attempts to replicate the experiment in which the guards were not directly instructed to abuse the prisoners, findings quite different (though no less interesting) from those of the Stanford prison experiment emerged.[20]

Despite the fact that the experiment falls far short of demonstrating its primary claimed conclusion, scientific textbooks continue to spread the conventional story about its significance. A detailed survey of seven contemporary social psychology textbooks written by experts in the field that included discussion of the Stanford prison experiment found that only two provided anything approaching a balanced discussion of the criticisms leveled against it. Closer to home, the same biased reporting is evident in discussions about the unconscious mind. In chapter 5 we described the implicit association test (IAT), a workhorse tool for purportedly measuring unconscious racial and other forms of bias, and some of the many criticisms leveled against this tool. A major concern is the paucity of evidence (despite many efforts to find such evidence) that IAT scores predict observable real-world behaviors indicative of bias. In an analysis of the way the IAT is discussed in seventeen introductory psychology textbooks, only two mentioned the dubious record of the IAT as a predictive tool.[21] It seems far more acceptable to textbook authors to tell a largely mythical story about psychological research than to give a more nuanced (but arguably more truthful) assessment. Discussing problems with a piece of research might muddle the story and create confusion in readers' minds. A good story, in contrast, may engage students and help to sell textbooks, but at the cost of misrepresenting reality.

Science does a poor job of correcting itself when initial eye-catching findings are later found to be either partially or wholly incorrect or are for some other reason discredited. Every year there continue to be numerous favorable citations to the research of the Dutch social psychologist Diederik Stapel, despite the fact that he fabricated data for his experiments (and admitted as much—as we saw in chapter 8). His studies have been formally retracted by the journals in which they were originally published but remain accessible. Despite the retractions, nontrivial numbers of scientists continue to discuss the findings of his research as if they have never been challenged.[22] Often this is presumably a consequence of lazy cutting-and-pasting when making a minor point that is not central to the scientist's research report, but it nonetheless highlights the fact that science itself struggles to ensure a balanced assessment of evidence, even in the most extreme and incontrovertible cases.

The Scientific Ecosystem

These problems with the ways in which scientists conduct their research are compounded by an ecosystem that encourages behaviors that are at variance with the pure pursuit of truth. Science is a competitive field. Scientists are employed by institutions that compete for students and prestige, they apply to competitive grant funding agencies to sponsor their research, and they submit their findings to journals that compete for subscribers and citations (when research articles include an earlier publication in their bibliographies, a standard currency for measuring the influence of the cited publication). Top scientists are headhunted at vast expense. At all of these stages, incentives are created that can pull the researcher away from the neutral pursuit of the truth.

If researchers are incentivized to try to maximize the number of publications they generate or the number of citations their research receives, then as sure as night follows day, they will modify their behavior to achieve these goals, even if it's at the expense of producing high-quality research. Goodhart's maxim tell us that when a measure becomes a target, it ceases to be a good target, and this is manifestly the case in the academic universe in which researchers are rewarded for the number of journal articles they produce and the citations they receive.[23]

One particularly stark but simple illustration that the quantity/quality balance is awry in much of psychological research is provided by surveys of the sizes of the samples researchers employ to test hypotheses in their experiments. This is a basic feature of research. After framing the hypothesis that she wishes to test and the measures and manipulations that will be employed to test it, the researcher must make decisions about the participants who will be tested—their age, characteristics, and, most important for this discussion, how many. It has been known for decades that these sample sizes tend to be too small. Imagine you're a researcher interested in measuring the effect of a particular intervention, say the effect of CBT on the symptoms of depression in primary care. In a standard randomized control trial (RCT), you might administer CBT to one group and a control or placebo treatment to another group. But how many people should be included in each group?

Past meta-analyses on this topic have found that the effect of CBT on depression—one of the best-established nonpharmacological treatments there is—has an effect size of about 0.2 in Cohen's *d* units.[24] Remember that the effect size for the male-female difference in height is about ten times this value, so against this benchmark, the beneficial effects of CBT are quite small, though of course when administered across many patients, this nonetheless amounts to a very meaningful therapeutic benefit. But our question is about statistical "power": How many people should the researcher include in each group of her RCT in order to be reasonably confident (say, 80 percent confident, which is the standard level adopted) of obtaining a statistically significant difference in measured depression symptoms between the two groups? The answer is that about 600 people are needed in total, assuming half are allocated to each group. This is a very large sample; the study would likely take many months to run and be a considerable time, money, and effort commitment for the researcher.

In reality, studies on the effects of CBT on depression in primary care, of which there have been over 30, have an average sample of about 160, vastly smaller than the sufficient size. Indeed this figure is in line with wider surveys of published research in psychology, suggesting that the average sample size is close to around 100 to 200,[25] and even this figure may wildly exaggerate the average for experimental psychological research, including studies on unconscious mental processes.[26] What this means is that researchers are often running experiments that are too weak to observe effects, even if those effects really exist. It may be the case that typical effects studied in psychological research in the field and laboratory are generally slightly bigger than that of CBT on depression, averaging a Cohen's *d* of, say, 0.4, but the sample sizes researchers use are still too small and are increasing at a glacial rate, if at all.[27]

Why do scientists tend to underpower their studies? The answer is not hard to discern: running smaller studies takes less time and resources and hence enables more articles to be published, leading to faster promotion, a bigger reputation, and so on. One might wonder what the point is of running an experiment with an inadequate sample size. Surely doing so raises the risk that the experiment will yield a nonsignificant, unpublishable result. This is where *p*-hacking comes to the rescue. Switching the outcome variable, for example, may exchange a statistically nonsignificant and unpublishable result for a significant and publishable one. And running underpowered studies doesn't simply increase the chances of wrongly obtaining null results; it also increases the likelihood that those studies that do, by good fortune, yield statistically significant results are false positives. As power decreases in a set of experiments, the ratio of false to true positives increases. It is not hard to see how inadequate sample sizes, resulting from the inherent pressures of the scientific ecosystem, can contribute to the creation of literatures like the money-priming one.

Individual researchers conducting their studies either independently or in collaborative groups represent one point on the pipeline for the generation of published articles. At other points along this pipeline, there are further incentives that can undermine the smooth and unbiased pursuit of truth. Even before investigators begin to collect data, funding agencies and industrial partners decide which projects to support, and it is well known that the funding source can bias the outcomes of the research. Large-scale meta-analyses reveal, for example, that drug trials funded by pharmaceutical companies yield more favorable outcomes than ones funded from other sources such as national research agencies.[28]

Closer to the end of the research pipeline are scholarly journals whose behavior also establishes perverse incentives that are often not aligned with the pursuit of truth. In the past, there tended to be a degree of commercial separation between a journal's publisher and its editorial team. In many cases, a journal would be strongly affiliated to a non-profit-making learned society whose purpose is to use journal revenue to fund researchers—particularly early-career scientists—working in its particular field, run research workshops and conferences, provide travel grants, and so on. Hence, the model involved universities paying a journal subscription to the publisher and the publisher handing over an agreed annual amount to the learned society for the privilege of having its badge of esteem on the journal. The society would provide the entire editorial team for the journal, deciding on its overall policy and making decisions on each manuscript submitted to it for evaluation.

It is easy to imagine that in such a model, editors have very little at stake other than the preservation of academic rigor, in whether any submitted manuscript is published. Editors, who by day are typically university employees, receive no remuneration for their work and certainly do not stand to gain financially from the publication of submitted manuscripts. But the idea that journals and their editors are disinterested referees solely concerned with maintaining scientific rigor is little more than an idealization. The reality is that even among reputable journals, the scientific ecosystem rewards behaviors that are not necessarily well aligned with the production of high-quality research. There is an understood hierarchy of journal prestige, with journals like *Science* and *Nature* at the very top. It can be a career-changing event for a young scientist to publish an article in one of these—as the fraudulent psychologist Diederik Stapel did in 2011. Usually this prestige is quantified by bibliometric indicators such as the journal's impact factor, which measures how frequently articles in that journal are cited by other researchers across a one-year period following publication. But there is no evidence that the research these journals publish is of higher quality than research published in more modest journals. On the contrary, there are many examples of psychology results published in *Science* and *Nature* proving unreplicable; the money-priming saga, for instance, would probably never have happened if the original report had not been published in *Science*, and other high-profile instances of unreplicable results relating to supposedly unconscious mental processes are easy to find.[29]

More worrying, there is even emerging evidence that some aspects of quality are inversely related to journal impact factor. The Center for Open Science has recently produced a ranking of science journals including many psychology ones according to the efforts they are making to promote transparency by requiring all articles to provide open data and materials, by supporting or even requiring preregistration, and so on. Against a maximum score of 30, both *Nature* and *Science* score a distinctly moderate 11. Another ranking system for journal quality, the *N*-pact factor,[30] ranks journals by one of the key factors we discussed earlier in this chapter: the average sample size of each experiment. The rationale is that everything else held equal, studies with higher statistical power are better ones, and hence a journal publishing such studies is fostering high-quality research. Evidence again suggests a minimal correspondence between journal impact factor and this quality index.[31]

Journal editors could easily and rapidly change the prevalent culture by requiring authors to adequately power, preregister, and replicate their experiments; make all their data and code openly available, and so on. But they fear that authors would go to competitors and their journal would lose market share and prestige. Editors are no different from other scientists in reacting to prevailing incentives. A particularly stark illustration is that some editors (thankfully a very small proportion) coerce authors, prior to accepting their submitted manuscripts for publication, to include in the bibliographies citations to articles previously published in that journal, for no other reason than to boost the journal's impact factor. Many journals and their editors (a rather larger proportion) are also immensely reluctant to publish corrections or retractions of articles shown to be faulty, presumably for fear of harming the journal's brand.[32]

Now fast-forward to the prevalent current publishing model, and we see that things may be getting even worse. In the new model—for excellent reasons to do with openness—publishers receive income not from subscriptions but from per article fees. Under this "open access" model, when a journal agrees to publish an article, the researcher or her university or grant funding agency pays a processing fee to the publisher, often over $1,000 (for the journal *Nature*, the eye-watering fee is over $10,000). On publication, the article is then publicly accessible by anyone, so the model achieves the laudable aim of making research universally available. The publisher justifies the fee by reference to the costs associated with editorial and production work, artwork, maintaining the digital repository, the cost of marketing the journal, and so on. Unfortunately, this model opens up an easy opportunity for unscrupulous businesses to launch "predatory" journals, which exist solely to make a profit and have no interest in academic rigor. Such journals bombard researchers with emails enticing them to submit their work, but in reality they carry out virtually no gatekeeping role in regard to standards, as is made abundantly clear by the many examples of such journals agreeing to publish hoax articles. One journal, for instance, published a deliberate hoax purporting to demonstrate that eating chocolate is a way of losing weight.[33]

The scientific ecosystem brings many forces to bear that support and promote poor-quality research. One effect of this has been to grossly distort our understanding of consciousness, although the consequences span probably the whole of science. Nevertheless, the growing recognition of these problems, the emergence of journals dedicated to fostering transparency, and the rapid increase in replications, preregistered or otherwise, over the past few years gives some reassurance that the culture in science is changing. Indeed surveys of economists, sociologists, psychologists, and political scientists confirm an emerging change toward the adoption of and support for practices designed to foster transparency, such as preregistration and making data, materials, and analysis code openly available. Whereas a minority of researchers adopted any of these practices fifteen years ago, far more do today.[34] But there remains a very long way to go.

11 The Mind Reclaimed?

On December 31, 2019, the Wuhan Municipal Health Commission informed the World Health Organization of some cases of "pneumonia of unknown etiology."[1] As is now common knowledge, these were the first recorded cases of COVID-19, a disease that has killed more than 6 million people and infected almost half a billion worldwide.[2] The impact of COVID-19 on society is difficult to overstate, not just for those of us who have been directly affected via illness and death of loved ones, but also via the changes to the way we work, live, socialize, and communicate. The pandemic has brought into sharp focus the urgent need to understand and influence human behavior. In many ways, it presents an opportunity for psychological and behavioral science to shine—to show their worth in helping us adapt to the "new normal" of living with COVID-19. In this final chapter, we use the pandemic to illustrate the importance of having well-developed and falsifiable theories of behavior if we are to use them to guide us, and we argue that research on unconscious thinking has largely failed to provide such theories. We also highlight that in the weak theories that have been put forward, unconscious mental processes seem to operate like dark matter in the universe—the residue that we infer "must" be there, but for which we have little direct evidence and even less theoretical understanding.

Promoting Social Distancing

A very early piece of health advice from governments around the world was to socially distance. In 2019, few of us would have known what that meant, but now the idea that we should maintain space between ourselves and others in order to stop the spread of the virus has become well established. If we want to understand why people may or may not comply with the request to socially distance then we need to have a theory of the psychological drivers of social distancing behavior. Absent a solid theory, our attempts to influence and maintain behavior changes will be futile.

In chapter 1 we discussed the theory of planned behavior (TPB); it turns out that along with its many successes across a broad spectrum of behaviors, it can be applied to understanding social-distancing compliance.[3] The TPB proposes that the extent to which people will socially distance depends on their *intention,* which in turn comes from three sources: their *attitudes* toward social distancing, their *subjective norms* regarding social distancing, and their *perceived behavioral control* over social distancing. Figure 11.1 shows how these aspects are related to one another. The important aspect of the theory, for our argument, is that behavior is determined by conscious knowledge, beliefs, and attitudes. There is no box or arrow in figure 11.1 for unconscious mental processes.

How does TPB fare when used to predict social distancing? To answer this question, in April 2020, when much of the world was under stay-at-home orders or lockdowns, Laura Gibson from the University of Colorado and colleagues asked a group of US adults about their attitudes, subjective norms, and perceived control toward social distancing. Attitudes were assessed via questionnaire items such as whether they found social distancing "healthy" or "unhealthy," subjective norms inquired about whether they thought most of their friends were engaging in social distancing, and perceived control measured confidence in being able to socially distance. For each item, a numerical rating was given, for instance from 1 (strongly disagree) to 7 (strongly agree). Participants then answered questions on their intentions to social distance over the coming weeks and months, and their actual distancing behavior over the preceding two weeks. In July 2020, when many lockdowns had been lifted, the same participants were approached and asked the same set of questions. This longitudinal design allows us to see how people's intentions, perceived control, and behavior are related across time.

A clear pattern emerged: the baseline measures, taken in the first survey in April, showed that attitudes, subjective norms, and perceived behavioral control predicted intentions to socially distance. Moreover, these intentions also predicted the extent of social distancing at follow-up in July. Finally, the level of perceived control measured in April was related to behavior in July, suggesting that control can bypass intentions to influence behavior. In

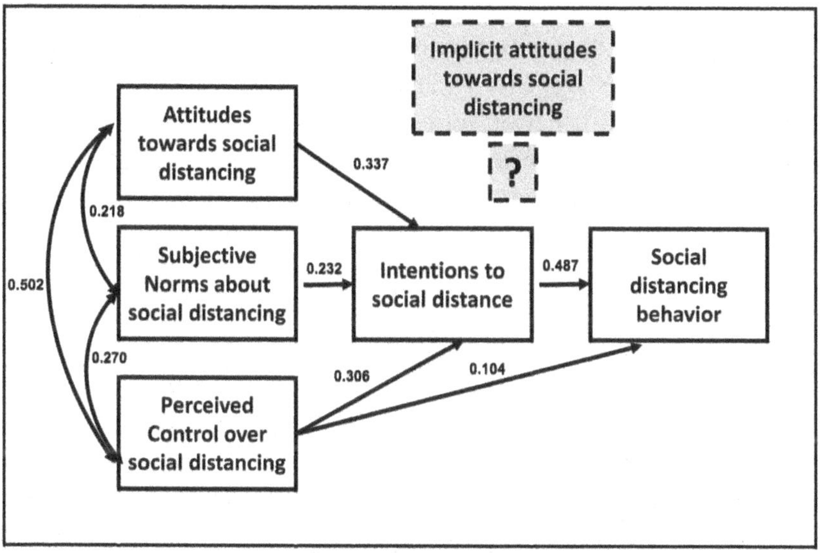

Figure 11.1
Theory of planned behavior applied to social distancing in the COVID-19 pandemic. Hundreds of studies have established that intentions to behave are driven by attitudes toward that behavior, subjective norms about those behaviors (what others do and think), and perceived control over engaging in the behavior. In turn, these intentions explain reported behaviors. The numbers indicate the strength of each link, where 1.0 is a perfect correlation and 0 is no relationship. The gray box marked "implicit attitudes" and a question mark illustrate uncertainty about how an implicit measure would be added to the theory. The data are from Laurel P. Gibson, Renee E. Magnan, Emily B. Kramer, and Angela D. Bryan, "Theory of Planned Behavior Analysis of Social Distancing during the Covid-19 Pandemic: Focusing on the Intention–Behavior Gap," *Annals of Behavioral Medicine* 55, no. 8 (2021): 805–12. https://doi.org/10.1093/abm/kaab041.

other words, the conscious, reportable beliefs and attitudes elicited by the questionnaires provide a good account of reported behavior. The numbers on the links in figure 11.1 indicate the strength of each link.

The appeal of such a clear set of results is that they provide direct guidance to policymakers wishing to develop health communications or interventions. Public health campaigns aimed at increasing social-distancing compliance should target attitudes, subjective norms, or perceptions of control (or some combination of these). Exactly how such targeting should be done is, of course, a question for another research program. One obvious extension to the Gibson study would be to use experimental designs (or randomized controlled trials) in which different groups of participants are presented with messages that are aimed to influence the different factors thought to underlie behavior. For example, one group might receive messages that emphasize subjective norms ("80 percent of people in your local area are socially distancing"), while another group see messages relating to perceived control ("going shopping early or late in the day helps avoid crowds"). Such designs would take us beyond the correlational findings and allow for causal inferences about the strength of different interventions.

One might ask whether the addition of unconscious factors into the TPB framework would improve its ability to predict and explain behavior. For example, one could imagine adding a box for "implicit attitudes" to figure 11.1 that is connected to intentions. The challenge then would be to demonstrate that the inclusion of such factors produced a better fit to the observed data or a better explanation of behavior than a model that omitted them. In essence, we'd be asking what the value-added, in terms of understanding behavior, is of assuming that some determinants of social distancing are unconscious. For example, maybe people's explicit attitudes to social distancing are less predictive than their implicit ones. People might say that social distancing is not unpleasant when asked directly, but if we could tap into their implicit attitudes, we might find a very different opinion. To do this, we need to measure those implicit attitudes, and that is not as straightforward as it might seem.

The Validation Crisis

The Implicit Association Test (IAT) discussed at length in chapter 5 provides a way to measure implicit attitudes. It is useful to return to this discussion because it provides an example of how one might test the idea that implicit attitudes are an additional predictor of behavior beyond explicit ones. Recall that the IAT attempts to measure automatic evaluations by combining two types of responses concurrently (see figure 5.1). For the purposes of illustration, let's imagine we want to measure unconscious ageism. In the experiment, you would see a sequence of randomly interspersed words and faces and have to press the left key for negative words and older faces and the right key for positive words and younger faces. If you have an unconscious bias against the elderly, it will be relatively easy to respond rapidly in these circumstances as the separate decisions are compatible in the sense that the left key is used for both negative words and the relatively disliked (older) faces and the right key for both positive words and the relatively liked (younger) faces. Your button presses should be faster on average under these circumstances than in another part of the experiment in which you have to press the left key for negative words and younger faces and the right key for positive words and older faces, where the separate decisions are incompatible. The difference between speed of responding in the compatible and incompatible stages is the IAT's estimate of your unconscious ageism.

Just as in the COVID-19 social-distancing example, we can build a theory in which unconscious ageism as measured by the IAT is assumed to cause some important aspect of decision making such as favoring younger compared to older job applicants. If we have IAT scores and job applicant ratings for a large sample of individuals, we can ask whether these measures are correlated. We can make our model more complex by adding a second potential cause of discriminatory behavior, namely, conscious or explicit ageist attitudes, which we could measure using a standard questionnaire (for example, the Expectations Regarding Aging scale).[4] Now we can ask whether unconscious, implicit attitudes (measured by the IAT) or conscious, explicit attitudes (measured by the questionnaire) correlate better with our measure of behavior.

This sounds relatively straightforward, but in reality, a great deal of further work is needed to validate our implicit attitude measure. Establishing the validity of a construct and the ability to measure it accurately is not trivial. Indeed, even though the notion of construct validity has a venerable history in psychology, it is often given insufficient attention in the development of our theories. This reticence to engage properly with the definition, refinement, and measurement of constructs has contributed substantially to the parlous state that the behavioral sciences now find themselves in.

To illustrate how concepts develop, consider an example from outside psychology. The concept "electron" was introduced to physics in the 1890s. Initially the term meant an elementary unit of charge. But over time, with experimentation, theoretical advances, and step changes in our understanding brought about by quantum theory, the meaning has changed radically. An electron now refers to an elementary particle that is a fundamental constituent of matter, with a negative charge of 1.602×10^{-19} coulombs, a mass of 9.108×10^{-31} kilograms, a spin of ½, and so on. Achieving such a level of precision in the definition of psychological concepts and linking them in precise ways to other theoretically related concepts might seem unattainable, but that is no reason to abandon our attempts to do so.[5]

One of the reasons construct validation is so challenging in psychology is that just as with electrons, we cannot directly observe mental processes. As we discussed in chapter 7, much of psychology and cognitive science is dedicated to inferring the impact of these latent mental states on behavior. We cannot observe perceived control, but we can draw inferences about the level of control someone has over a particular behavior. This can be done either by asking them directly or by measuring the impact on behavior of a manipulation that we think will affect their control. This is the approach that has been taken in hundreds of studies testing TPB and has enabled us to build confidence in the model depicted in figure 11.1. But these attempts at measurement depend fundamentally on our assumptions that our tools measure what they are intended to measure. This is the essence of establishing construct validity.

Returning to our ageism IAT example, let's suppose that our findings suggest that implicit but not explicit attitudes are strongly associated with behavior. Before concluding that our behavior toward other people is strongly influenced by unconscious ageist attitudes, we need reassurance that we measured conscious attitudes in a sound manner. Perhaps our questionnaire items are simply not very appropriate and fail to sensitively discriminate between people with and without ageist attitudes. Or perhaps our questionnaire is insensitive because the respondents were wary about revealing their true views. There are any number of ways in which a test may be a poor assessment of the construct it is designed to measure. Note that validity is not quite the same as exhaustiveness. In chapter 2 (see table 2.1) we discussed some of the criteria that adequate tests of awareness need to meet before we can deem them sufficiently exhaustive and sensitive to be usable; for instance, awareness tests should be administered at or very close to the time at which behavior is being measured. Validity includes exhaustiveness but encompasses other attributes too, such as convergence with other tests designed to measure the same construct (awareness in this case).

Against the standard criteria for validation, it is debatable whether the IAT provides a good measure of unconscious attitudes.[6] Primarily this is because its incremental predictive validity—the extent to which it predicts behavior over and above consciously reportable attitudes—is tiny.[7] But this is not the key point. At least with respect to the IAT, considerable efforts have been made to validate it as a measure of unconscious attitudes, and its properties have been fairly thoroughly assessed. When we turn to other domains in which unconscious processes have been investigated, we find that the picture regarding the IAT is very much the exception. In research on subliminal perception, priming, decision making, and many other areas, virtually no efforts have been made to establish that the constructs being measured have any validity. There is, in short, very little reason to believe that something like money priming (that we reviewed extensively in chapters 9 and 10) can be measured validly and distinguished from conscious attitudes. Despite hundreds of publications on money priming, the relevant research needed to establish its soundness as a measurable and distinct psychological construct has simply not been undertaken.

The consequences of inadequate construct validation are hard to exaggerate. Different measurements, superficially probing the same construct, can yield completely contrasting results. In a major investigation led by Justin Landy, teams of investigators were asked to devise their own methods for answering several different hypotheses, one of which related to unconscious thoughts:[8]

Hypothesis: People explicitly self-report an awareness of harboring negative automatic associations with members of negatively stereotyped social groups.

This hypothesis conjectures that people tend to have some insight into their own biases and prejudices against certain social groups such as the elderly or racial minorities. On the face of it, the hypothesis looks as if it should be fairly easy to test. A sample of randomly chosen individuals could be given a question such as

Q1: Although I don't necessarily agree with them, I sometimes have prejudiced feelings (like gut reactions or spontaneous thoughts) that I don't feel I can prevent,

and asked to what extent they agree or disagree using a scale from 1 ("strongly disagree") to 7 ("strongly agree"), and with 4 ("neither agree nor disagree") being the crucial midpoint. We could then ask whether the average rating on this scale is greater than 4, indicating that indeed most people explicitly self-report an awareness of harboring negative automatic associations with members of negatively stereotyped social groups, or instead is close to or even below 4, in which case the hypothesis is falsified.

Suppose we obtain the former outcome (which was indeed the case): Does that mean the hypothesis is correct? Bear in mind that although our question (Q1) seems intuitively to probe beliefs about the construct we're interested in, we could equally measure those same beliefs in a virtually infinite number of other ways. Without undertaking a validation exercise, we have no way of knowing whether these different ways would all yield similar results and we don't actually know that Q1 probes the target construct at all. If we find an average rating of, say, 5 in response to our question, what does that tell us?

The results of Landy's project were stark: different teams, constructing different questions to test the same hypothesis, reached completely divergent conclusions. For example, one team employed this alternative question wording:

Q2: Regardless of my explicit (i.e., conscious) beliefs about social equality, I believe I possess automatic (i.e., unconscious) negative associations towards members of stigmatized social groups.

With this wording, the majority of respondents did *not* report harboring negative automatic associations. Hence two questions, both of which seem superficially to get at the same psychological state, yield quite contrasting results. Indeed across the many teams that participated in Landy's project, this was the typical pattern found for all of the research hypotheses investigated. Each team constructed (on the face of it) completely reasonable questions to test the same hypothesis, but the conclusions reached by the various teams did not converge. The message is plain: it is not sufficient to appeal to plausibility when constructing test items; they need to be subjected to a thorough validation exercise. If we want to determine whether Q1 or Q2 or indeed any other question provides a valid means of measuring beliefs about negative automatic associations, we need to establish convergent validity (the question should yield scores that are similar to other questions designed to measure the same beliefs), discriminant validity (the question should yield scores that are unrelated to the scores for questions designed to measure different beliefs), and predictive validity (it should predict some meaningful aspect of behavior that we expect to be related to beliefs about negative automatic associations, such as willingness to attend an unconscious bias training course). The question should also be reliable, in the statistical sense of yielding roughly similar scores when any given individual answers the question on different occasions.

This neglect of validation is widespread across psychological science to such an extent that it has been described as a "validation crisis."[9] Instead of carefully measuring and characterizing latent psychological states, beliefs, or dispositions, researchers focus unduly on very simplistic questions such as, "Does an increase in A cause an increase in B?," answering such questions solely by reference to statistical significance and the $p < .05$ threshold.

Theory Building and Severe Tests

There is an important distinction to be made here though. Some phenomena (for example some forms of priming, facial feedback effects) have proven to be very hard to replicate, leading the research community to be skeptical about their existence. This contrasts with phenomena that are highly robust, but for which explanations for their properties are still hampered by poor construct validation. Take the example of the bat-and-ball problem presented at the start of chapter 7 (if a bat and a ball cost $1.10 in total and the bat costs $1 more than the ball, how much does the ball cost?). It is undisputed that the majority of people get this problem wrong when they first see it, and this behavioral regularity requires an explanation. The standard flavor of this explanation follows the dual-system framework depicted in figure 7.1. People fail to engage their deliberative system 2 processes, and so the automatic system 1 provides a plausible but incorrect answer. Performance on the bat-and-ball problem along with two similar questions is then purported to measure the degree to which people can inhibit system 1 when they should be listening to system 2, a construct described as *cognitive reflection*.

But is cognitive reflection an ability or a disposition to think in particular ways? And if it is an ability, how does it relate to other general cognitive abilities such as intelligence or working memory? Some progress has been made toward answering these questions with the general consensus that the Cognitive Reflection Test (CRT) measures something more than simple mathematical or reasoning ability.[10] The test also appears to have incremental predictive validity over similar measures in predicting other types of behavior such as people's willingness to gamble. But this is not the main problem. Even if one accepts that cognitive reflection is an independent psychological construct that is measured validly by the CRT, the broader question of how it relates to all of the other constructs and capabilities listed in figure 7.1 remains unanswered.

The dual-system framework outlined in figure 7.1 is a far cry from the detailed *nomological net*—a term used to describe how the different constructs in a theory are related to one another—shown in the TPB COVID-19 example of figure 11.1. The construction of these nets is the hard but fundamentally important part of theory development. Without clear statements of how and why our different, validated, psychological constructs interrelate, we remain in a quagmire of loose relationships with untestable and unfalsifiable theories. As we noted in chapter 7, many of the dual-system models include collections of different constructs—capacity and automaticity—that are assigned to different systems with little attempt to establish a coherent understanding of how these constructs link to one another or indeed whether the measurements of them are valid. To reiterate our conclusion from chapter 7, the result of this incoherence is that the attempts to develop dual-system theories become akin to the epicycles devised by Greek astronomers to rescue the theory that the Earth revolves around the sun; their contact with hard facts becomes more and more distant while precise predictions that could be subjected to experimental tests are scarce.

At the start of this book, we suggested that the science of psychology needs to be rebuilt from the ground up on firmer foundations. Those firm foundations are our theories of behavior. So why is theory building so difficult? To give a sense of the challenges, imagine that we propose a theory that could explain some of the more surprising findings that we've reviewed in earlier chapters. What kind of theory might explain why we'd walk more slowly down a corridor after solving anagrams related to the concept of old age, or rate cartoons as funnier when we hold a pen between our teeth, or rate someone as warmer on a personality scale after holding a hot cup of coffee rather than a cold can of Coke? We now know that many of these findings do not replicate, and thus the need to find explanations becomes moot, but for a moment, let's consider a world in which these or similar findings were real. How might we explain them?

We might propose a general theory called "embodiment priming theory."[11] The core assumptions of this theory are as follows. First, abstract concepts like warmness toward an individual are grounded in bodily states, sensations, or movements, like the physical sensation of warmth. Second, inducing that bodily state or sensation, for instance, by giving someone a warm cup, activates or primes that concept. And finally, the concept influences downstream behavior related to that concept, such as rating the warmth of a stranger. The tenets of the theory are similar to the ideas we explored in chapter 3's discussions of the ripples of activation. How would we test embodiment priming theory? The first challenge is defining the scope of the theory or the space of potential hypotheses that we might want to test. There are many abstract concepts that can be grounded in innumerable bodily states and sensations that can be induced experimentally in a host of different ways. There is also a variety of downstream behaviors that could be influenced by the primed concept. Hypotheses are limited only by the ingenuity of the experimenter. For example, we might test the hypothesis that a person in the UK who stands on their head for three minutes will feel more connected to people in Australia by priming the concept of being Down Under.

The problem here is that the combinatorial explosion that results from pairing concepts, groundings, their manipulations and measurements presents an asymmetry in tests of the theory. Let's imagine we ran the Down Under experiment and found no effect. Does that falsify the theory? Not necessarily; we could argue that we chose the wrong amount of time for standing on your head—perhaps 1 minute is better because after 3 minutes, people become woozy and don't feel connected to anyone. Or perhaps it only works for Brits who have relatives in Australia. There are many ways to argue ourselves out of a failure to find an effect but to remain convinced that the embodiment priming theory is a space of hypotheses worth exploring. Of course, if we are lucky enough to find the effect (perhaps with the aid of *p*-hacking), then our confidence in the general applicability of the theory increases. But should it?

The answer lies in how tightly our hypotheses are linked to our theories. If a theory strongly implies a given hypothesis, then attempts to test those hypotheses are useful in the sense that they are diagnostic about the boundaries of the theory. This brings us back to the crucial importance of construct validity. If the elements in our theory are loosely defined or poorly validated, then any tests of the theory will be hamstrung from the start. If we do not have a well-validated concept of Down-Underness, then finding an effect of standing on our heads will be meaningless. A useful distinction here is between discovery-oriented and theory-testing research (we met similar ideas before in chapter 9 when discussing exploratory and confirmatory research). Discovery-oriented research is what we are doing when we are searching, perhaps stumbling, through the space of hypotheses that our theory implies. It describes situations where we focus on testing a single hypothesis with a single manipulation and care less about how that collection of effects (and their absences) integrates into a broader understanding. Theory-testing research, in contrast, is where we focus on strengthening the inferential links between well-established empirical findings and formalized theories that can explain them.

As we have argued throughout the preceding chapters, psychology has been stuck for a long time in cycles of discovery-oriented research, and much of this has been fueled by increasingly implausible claims about the power of unconscious mental processes. Most frustrating, the unconscious is treated in many of these explanations like dark matter in the universe—the residue that we infer must be there, but for which we have no theoretical understanding. To make real progress, we must abandon these black-box explanations and construct genuinely testable explanations of human behavior.

A strong theory-led approach, combined with the improvements in research practices that we've discussed in this book, is the way forward to a true science of human behavior. A piecemeal approach in which we address one aspect but not the other will be insufficient. For example, a pure focus on improving methods will not be enough to strengthen theories. The recent enthusiasm for preregistration provides a good illustration. The act of preregistering the hypothesis that standing on your head in the UK makes you feel more connected to people in Australia does not in and of itself make it a stronger or better hypothesis. Moreover, if a researcher happened to find an effect that supported the hypothesis, it should not necessarily increase its credibility. What makes a hypothesis credible is how closely it is connected to or constrained by the theory that motivates it. Without a detailed nomological net outlining how our (validated) construct of Down-Underness relates to the physiological state induced by being inverted, which in turn influences our feelings about distant others, a statistically significant effect is no more informative than a lucky guess. And preregistering that lucky guess won't help.

This is not to say that preregistration is without merit. As discussed in chapters 9 and 10, it can help to negate the impact of HARKing (hypothesizing after the results are known) and forces researchers to be explicit about how, why, and how strongly their predictions are justified by their theory before seeing any data. But absent the strong theory to begin with, preregistration appears to be best thought of as a cure for the symptoms of our current malaise than a solution for the core problems.[12]

One way that our predictions could be made more precise and subject to severe and decisive tests is to use the tools of computational modeling. In addition to wanting to understand the magnitude of relationships between well-measured constructs, we also often want to know what exactly the mental machinery is and what the algorithms are that implement these relationships. Computational modeling, in the form of computer programs designed to capture the essence of these algorithms, plays an important role in addressing such questions.[13] We might speculate, for instance, that the process of choosing between two options, A and B, involves an evidence-accumulation process in which, moment by moment, internal counters accumulate reasons for choosing each option, and a decision is made as soon as one of these counters reaches a prespecified threshold. Such a mechanism could be turned into a computer model and used to generate testable predictions about decision making. Computational models have made an enormous contribution to modern psychological science, yielding deep mechanistic explanations for decision making, perception, memory, and other cognitive processes. Usually the aim is to compare two or more models to a rich set of data to see which fits the data more closely rather than to focus on a single question about whether an effect is or is not statistically significant. Arguably, the best way to test verbal theories is in quantitative terms—that is, through models.

As with construct validation, however, the computational approach has been underemployed in the study of unconscious mental processes. The general approach would be to construct a computational model of whatever aspect of behavior is under investigation (response times or choices, for example) and then compare a model that includes an unconscious aspect with one that does not to gauge whether the former appreciably improves explanatory power. In the rare cases where this approach has been taken, little evidence for such an outcome has been found, but again that is not the main point.[14] Rather, the key issue is the disappointingly limited use of the approach. When a well-validated tool is available to help us answer crucial questions about human behavior, it is depressing that the tool is not used.

This does not mean that computational approaches are a panacea. A construct does not necessarily become more valid simply because it is instantiated in a formal mathematical model. Neither is a theory that lacks formalization necessarily bad. A case in point is Darwin's theory of evolution. The theory has very clearly defined concepts or constructs, but the relationships between them remain qualitative rather than quantitative.[15] The central point remains, though, that for our understanding of human behavior to improve, we must not shy away from the difficult task of theory building.

Opening Our Science and Minds

This book is about the need to be open about our science, but more fundamentally, it is about opening our minds to what on the face of it seems like an uncontroversial idea: we are the authors of our own actions, we typically have good intuitions and understanding about the reasons for why we behave as we do, and thus we also have the ability to change our behavior. The fact that this idea brooks any controversy is testament to the reach of the powerful unconscious mind meme. Realizing that the notion of unconscious thinking is built on very shaky foundations has significant implications for how we lead our lives. In one sense, it should be deeply empowering: we are not constantly being buffeted by forces beyond our control, nor do we have deeply rooted unconscious biases and prejudices lurking at the bottom of the iceberg. But in another sense, this realization is unsettling: we now have to take responsibility for our failures rather than blaming them on our "brain" acting unconsciously. We also have to acknowledge that our prejudices are often conscious—in the sense of being accessible to our awareness—even if we seek to deny their existence or lay the blame elsewhere.

Just over one hundred years ago, the world was in the grip of another pandemic, the influenza pandemic that killed at least 50 million people and infected almost a third of the world's population.[16] George Soper, a major in the Sanitary Corps of the US Army, wrote an article at the time for the journal *Science*, "The Lessons of the Pandemic."[17] In a telling line, he noted that "the measures which were introduced for the control of the pandemic were based upon the slenderest of theories."[18] How far have we come in the intervening century? How much fatter have our theories become?

Soper argued that successful prevention required overcoming the public's indifference or underappreciation of the risks of spreading the disease. He suggested that much of the transmission occurred "unconsciously, invisibly, unsuspectingly." Soper's recommendations for the most essential behaviors to stop the spread are eerily reminiscent of the much-repeated slogans of the COVID-19 era: avoid needless crowding, smother coughs and sneezes, wash hands before eating, and open windows when practicable. Has a century's worth of psychological research brought us any closer to understanding how to achieve those changes in behavior? At the onset of the current pandemic, many researchers were swift to point out a raft of relevant findings from psychology that could help to mitigate the impact of COVID-19.[19] These ranged from research on basic science communication to work on leadership, stress, threat perception, and social context. There is no doubt that many of these studies could be applied to understanding and changing behavior. But taken as a whole, how much do they contribute to genuinely deep theory development? Do they constitute real progress toward achieving a true science of the mind, or are they an ad hoc collection of effects found in underpowered studies, using inappropriate methods and questionable statistical approaches? The answer must lie between these extremes. It would be churlish to argue that we've made no progress (take TPB for example) but naive to claim that our theories have allowed us to advance much beyond George Soper's recommendations. If we don't want to find ourselves in the same position in another one hundred years, then we must address the challenges outlined in this book. Abandoning the myth of the smart unconscious is a good place to start.

Notes

Chapter 1

1. "An Extraordinary Iceberg Is Gone, but Not Forgotten," *New York Times,* January 26, 2022, https://www.nytimes.com/2022/01/26/climate/iceberg-a68a-antarctica.html.

2. Interestingly, although Freud thought most mental processes take place "beneath the surface," the famous quote that opens this chapter is apparently incorrectly attributed to him. For details, see "10 Quotes Wrongly Attributed to Sigmund Freud," accessed April 27, 2022, at https://www.freud.org.uk/2019/04/30/10-quotes-wrongly-attributed-to-sigmund-freud/.

3. For example, see Timothy D. Wilson, *Strangers to Ourselves: Discovering the Adaptive Unconscious* (Cambridge, MA: Belknap Press, 2002).

4. Statements of this perspective can be found in George A. Miller, "Some Psychological Studies of Grammar," *American Psychologist* 17, no. 11 (1962): 748–762, https://doi.org/10.1037/h0044708; and Nick Chater, *The Mind Is Flat: The Remarkable Shallowness of the Improvising Brain* (New Haven: Yale University Press, 2018).

5. For example, see Jonathan St. B. T. Evans, "Dual-Processing Accounts of Reasoning, Judgment and Social Cognition," *Annual Review of Psychology* 59 (2008): 255–278, https://doi.org/10.1146/annurev.psych.59.103006.093629; Daniel Kahneman, *Thinking, Fast and Slow* (New York: Farrar, Straus and Giroux, 2011); David Halpern, *Inside the Nudge Unit: How Small Changes Can Make a Big Difference* (London: W. H. Allen, 2015).

6. This point is exemplified by the biostatistician John Ioannidis's startling claim, made not remotely in jest, that most published research is wrong: John P. A. Ioannidis, "Why Most Published Research Findings Are False," *PLOS Medicine* 2, no. 8 (2005): e124, https://doi.org/10.1371/journal.pmed.0020124.

7. Scott O. Lilienfeld and Irwin D. Waldman, eds., *Psychological Science under Scrutiny: Recent Challenges and Proposed Solutions* (Chichester: Wiley, 2017).

8. Richard H. Thaler and Cass R. Sunstein, *Nudge: Improving Decisions about Health, Wealth, and Happiness* (New Haven: Yale University Press, 2008).

9. See Jonathan Birch, Alexandra K. Schnell, and Nicola S. Clayton, "Dimensions of Animal Consciousness." *Trends in Cognitive Sciences* 24, no. 10 (2020): 789–801, https://doi.org/10.1016/j.tics.2020.07.007.

10. For an illuminating discussion of reports and how they have been treated by various schools of thought in psychology, see Eddy A. Nahmias, "Verbal Reports on the Contents of Consciousness: Reconsidering Introspectionist Methodology," *Psyche: An Interdisciplinary Journal of Research on Consciousness* 8 (2002).

11. The argument is made, for instance, in John A. Bargh and Ran R. Hassin, "Human Unconscious Processes *in Situ*: The Kind of Awareness That Really Matters," in *The Cognitive Unconscious*, ed. Arthur Reber and Rhianon Allen (New York: Oxford University Press, 2022). Bargh and Hassin say, "Much preparatory work needs to be done in order to create our conscious experiences, and by logical necessity, all this work is unconscious."

12. The representational theory of mind, which identifies mental states as ones that have representational content, is most strongly associated with Jerry Fodor. See Jerry A. Fodor, *Representations: Philosophical Essays on the Foundations of Cognitive Science* (Cambridge, MA: MIT Press, 1983).

13. See Daniel B. Rubin and Angelique C. Paulk, "Neuron, Control Thyself!" *Brain* 144, no. 12 (2021): 3550–3551, https://doi.org/10.1093/brain/awab413.

14. Probably the most famous critique of Freud's work is Hans J. Eysenck, *Decline and Fall of the Freudian Empire* (New York: Viking Press, 1985).

15. Mark C. Fox, Karl Anders Ericsson, and Ryan Best, "Do Procedures for Verbal Reporting of Thinking Have to Be Reactive? A Meta-Analysis and Recommendations for Best Reporting Methods," *Psychological Bulletin* 137, no. 2 (2011): 316–344, https://doi.org/10.1037/a0021663.

16. See Martin Fishbein and Icek Ajzen, *Predicting and Changing Behavior: The Reasoned Action Approach* (New York: Routledge, 2010).

17. See Gabriela Topa and Juan Antonio Moriano, "Theory of Planned Behavior and Smoking: Meta-Analysis and SEM Model," *Substance Abuse and Rehabilitation* 1 (2010): 23–33, https://doi.org/10.2147/SAR.S15168.

18. Fishbein and Ajzen, *Predicting and Changing Behavior.*

19. For one example of the possible value of adding unconscious processes, see Daniel J. Phipps, Thomas E. Hannan, Ryan E. Rhodes, and Kyra Hamilton, "A Dual-Process Model of Affective and Instrumental Attitudes in Predicting Physical Activity," *Psychology of Sport and Exercise* 54 (2021): 101899, https://doi.org/10.1016/j.psychsport.2022.102222.

20. Nagel's article is one of the most influential contributions to the philosophy of consciousness: Thomas Nagel, "What Is It Like to Be a Bat?" *Philosophical Review* 83, no. 4 (1974): 435–450, https://doi.org/10.2307/2183914.

21. The quotation is from Thomas H. Huxley, *Lessons in Elementary Physiology* (London: Macmillan, 1866), 193.

22. See Giles S. Brindley and Walpole S. Lewin, "The Sensations Produced by Electrical Stimulation of the Visual Cortex," *Journal of Physiology* 196, no. 2 (1968): 479–493, https://doi.org/10.1113/jphysiol.1968.sp008519. Interestingly, this demonstration of stimulation-induced phosphenes was in a blind individual.

23. See John-Dylan Haynes and Geraint Rees, "Decoding Mental States from Brain Activity in Humans," *Nature Reviews Neuroscience* 7 (2006): 523–534, https://doi.org/10.1038/nrn1931.

24. This demonstration, verging on science fiction but subsequently corroborated in other patients, was reported by Adrian M. Owen, Martin R. Coleman, Melanie Boly, Matthew H. Davis, Steven Laureys, and John D. Pickard, "Detecting Awareness in the Vegetative State," *Science* 313, no. 5792 (2006): 1402–1402, https://doi.org/10.1126/science.1130197.

25. The theory—particularly in neuroscience discussions—is often also referred to by a slightly different name: global neuronal workspace theory. Key statements of the theory can be found in Bernard J. Baars, "Global Workspace Theory of Consciousness: Toward a Cognitive Neuroscience of Human Experience," *Progress in Brain Research* 150 (2005): 45–53, https://doi.org/10.1016/S0079-6123(05)50004-9; and Stanislas Dehaene and Jean-Pierre Changeux, "Experimental and Theoretical Approaches to Conscious Processing," *Neuron* 70, no. 2 (2011): 200–227, https://doi.org/10.1016/j.neuron.2011.03.018. Recent work has enriched the theory by linking it to deep learning models from artificial intelligence. See Rufin VanRullen and Ryota Kanai, "Deep Learning and the Global Workspace Theory," *Trends in Neurosciences* 44, no. 9 (2021): 692–704, https://doi.org/10.1016/j.tins.2021.04.005.

26. Itay Yaron, Lucia Melloni, Michael Pitts, and Liad Mudrik, "The ConTraSt Database for Analysing and Comparing Empirical Studies of Consciousness Theories," *Nature Human Behaviour* (2022), https://doi.org/10.1038/s41562-021-01284-5.

Chapter 2

1. This pioneering research was originally reported in Benjamin Libet, Curtis A. Gleason, Elwood W. Wright, and Dennis K. Pearl, "Time of Conscious Intention to Act in Relation to Onset of Cerebral-Activity (Readiness Potential): The Unconscious Initiation of a Freely Voluntary Act," *Brain* 106, no. 3 (1983): 623–642, https://doi.org/10.1093/brain/106.3.623; see also Benjamin Libet, "Unconscious Cerebral Initiative and the Role of Conscious Will in Voluntary Action," *Behavioral and Brain Sciences* 8, no. 4 (1985): 529–539, https://doi.org/10.1017/S0140525X00044903.

2. The meta-analysis was undertaken by Moritz N. J. Braun, Janet Wessler, and Malte Friese, "A Meta-Analysis of Libet-Style Experiments," *Neuroscience & Biobehavioral*

Reviews 128 (2021): 182–198, https://doi.org/10.1016/j.neubiorev.2021.06.018. It is rather salutary that, in Braun et al.'s own words (p. 194), "For this difference, we could procure data from $k=6$ studies . . . based on only $N=53$ participants (!). This meager data base stands in sharp contrast to the huge influence the Libet experiment has had on both scientific thinking in various disciplines and public discourse." Across the entire forty-year span since Libet's introduction of this technique for measuring the subjective and objective timing of simple movements, only data from fifty or so individuals have ever been reported. Of relevance to issues that will become of central importance in later chapters, it does, however, include data from eight participants examined in a painstaking and largely successful replication of Libet's methods, with the full data being publicly accessible, by Tomáš Dominik, Daniel Dostál, Martin Zielina, Jan Šmahaj, Zuzana Sedláčková, and Roman Procházka, "Libet's Experiment: A Complex Replication," *Consciousness and Cognition* 65 (2018): 1–26, https://doi.org/10.1016/j.concog.2018.07.004.

3. Itzhak Fried, Roy Mukamel, and Gabriel Kreiman, "Internally Generated Preactivation of Single Neurons in Human Medial Frontal Cortex Predicts Volition," *Neuron* 69, no. 3 (2011): 548–562, https://doi.org/10.1016/j.neuron.2010.11.045.

4. Chun Siong Soon, Marcell Brass, Hans-Jochen Heinze, and John-Dylan Haynes, "Unconscious Determinants of Free Decisions in the Human Brain," *Nature Neuroscience* 11 (2008): 543–545, https://doi.org/10.1038/nn.2112.

5. The arguments reviewed in this section are based largely on the work of Jeff Miller and his colleagues. See Judy Trevena and Jeff Miller, "Cortical Movement Preparation before and after a Conscious Decision to Move," *Consciousness and Cognition* 11, no. 2 (2002): 162–190, https://doi.org/10.1006/ccog.2002.0548; Judy Trevena and Jeff Miller, "Brain Preparation before a Voluntary Action: Evidence against Unconscious Movement Initiation," *Consciousness and Cognition* 19, no. 1 (2010): 447–456, https://doi.org/10.1016/j.concog.2009.08.006; and Jeff Miller, Peter Shepherdson, and Judy Trevena, "Effects of Clock Monitoring on Electroencephalographic Activity: Is Unconscious Movement Initiation an Artifact of the Clock?" *Psychological Science* 22, no. 1 (2011): 103–109, https://doi.org/10.1177/0956797610391100. Recent reviews of the field were published by Marcel Brass, Ariel Furstenberg, and Alfred R. Mele, "Why Neuroscience Does Not Disprove Free Will," *Neuroscience & Biobehavioral Reviews* 102 (2019): 251–263, https://doi.org/10.1016/j.neubiorev.2019.04.024, and Edward J. Neafsey, "Conscious Intention and Human Action: Review of the Rise and Fall of the Readiness Potential and Libet's Clock," *Consciousness and Cognition* 94 (2021), https://doi.org/10.1016/j.concog.2021.103171.

6. This finding is reported in Uri Moaz, Gideon Yaffe, Christof Koch, and Liad Mudrik, "Neural Precursors of Decisions That Matter: An ERP Study of Deliberate and Arbitrary Choice," *eLife* 8 (2019): 39787, https://doi.org/10.7554/eLife.39787.001.

7. See William P. Banks and Eve A. Isham, "We Infer Rather Than Perceive the Moment We Decided to Act," *Psychological Science* 20, no. 1 (2009): 17–21, https://doi.org/10.1111/j.1467-9280.2008.02254.x.

8. Jeff Miller, Paula Vieweg, Nicolas Kruize, and Belinda McLea, "Subjective Reports of Stimulus, Response, and Decision Times in Speeded Tasks: How Accurate Are Decision Time Reports?" *Consciousness and Cognition* 19, no. 4 (2010): 1013–1036, https://doi.org/10.1016/j.concog.2010.06.001.

9. Libet, "Unconscious Cerebral Initiative," 529.

10. Many models of the gradual accumulation of neural activity prior to a movement have been proposed—for instance, by Aaron Schurger, Jacobo D. Sitt, and Stanislas Dehaene, "An Accumulator Model for Spontaneous Neural Activity Prior to Self-Initiated Movement," *Proceedings of the National Academy of Sciences* 109, no. 42 (2012): E2904–E13, https://doi.org/10.1073/pnas.1210467109; and Jeff Miller and Wolf Schwarz, "Brain Signals Do Not Demonstrate Unconscious Decision Making: An Interpretation Based on Graded Conscious Awareness," *Consciousness and Cognition* 24 (2014): 12–21, https://doi.org/10.1016/j.concog.2013.12.004.

11. See Sven Walter, "Willusionism, Epiphenomenalism, and the Feeling of Conscious Will," *Synthese* 191 (2014): 2215–2238, https://doi.org/10.1007/s11229-013-0393-y.

12. See his influential and thought-provoking book, Daniel M. Wegner, *The Illusion of Conscious Will* (Cambridge, MA: MIT Press, 2002).

13. See Albert Michotte, *The Perception of Causality* (London: Routledge, 1963); Tom L. Beauchamp and Alexander Rosenberg, *Hume and the Problem of Causation* (Oxford: Oxford University Press, 1981).

14. The experiment is described in Daniel M. Wegner and Thalia Wheatley, "Apparent Mental Causation: Sources of the Experience of Will," *American Psychologist* 54, no. 7 (1999): 480–492, https://doi.org/10.1037/0003-066X.54.7.480.

15. See Daniel M. Wegner, Betsy Sparrow, and Lea Winerman, "Vicarious Agency: Experiencing Control over the Movements of Others," *Journal of Personality and Social Psychology* 86, no. 6 (2004): 838–948, https://doi.org/10.1037/0022-3514.86.6.838.

16. See Walter, "Willusionism," for a review.

17. Eddy Nahmias, "Agency, Authorship, and Illusion," *Consciousness and Cognition* 14, no. 4 (2005): 771–785, https://doi.org/10.1016/j.concog.2005.07.002.

18. John McClure, "Attributions, Causes, and Actions: Is the Consciousness of Will a Perceptual Illusion?" *Theory & Psychology* 22, no. 4 (2012): 402–419, https://doi.org/10.1177/0959354310386845.

19. Rom Harré and Edward H. Madden, *Causal Powers* (Oxford: Blackwell, 1975).

20. This intriguing effect was reported by Petter Johansson, Lars Hall, Sverker Sikström, and Andreas Olsson, "Failure to Detect Mismatches between Intention and Outcome in a Simple Decision Task," *Science* 310, no. 5745 (2005): 116–219, https://doi.org/10.1126/science.1111709; Petter Johansson, Lars Hall, and Sverker Sikström,

"From Change Blindness to Choice Blindness," *Psychologia* 51, no. 2 (2008): 142–155, https://doi.org/10.2117/psysoc.2008.142.

21. Its more technical name is *optic acceleration cancellation theory*. For discussion of this theory, see Seville Chapman, "Catching a Baseball," *American Journal of Physics* 36, no. 10 (1968): 868–870, https://doi.org/10.1119/1.1974297; Zoltan Dienes and Peter McLeod, "How to Catch a Cricket Ball," *Perception* 22, no. 12 (1993): 1427–1439, https://doi.org/10.1068/p221427; Claire F. Michaels and Raoul R. D. Oudejans, "The Optics and Actions of Catching Fly Balls: Zeroing Out Optical Acceleration," *Ecological Psychology* 4, no. 4 (1992): 199–222, https://doi.org/10.1207/s15326969eco0404_1; Dees B. W. Postma, Joanne Smith, Gert-Jan Pepping, Steven van Andel, and Frank T. J. M. Zaal, "When a Fly Ball Is Out of Reach: Catchability Judgments Are Not Based on Optical Acceleration Cancelation," *Frontiers in Psychology* 8 (2017): 535, https://doi.org/10.3389/fpsyg.2017.00535.

22. Though see Postma et al., "When a Fly Ball."

23. This careful research is reported in Nick Reed, Peter McLeod, and Zoltan Dienes, "Implicit Knowledge and Motor Skill: What People Who Know How to Catch Don't Know," *Consciousness and Cognition* 19, no. 1 (2010): 63–76, https://doi.org/10.1016/j.concog.2009.07.006.

24. The concept of attribute substitution is elaborated by Daniel Kahneman and Shane Frederick, "Representativeness Revisited: Attribute Substitution in Intuitive Judgment," in *Heuristics and Biases: The Psychology of Intuitive Judgment*, ed. Thomas Gilovich, Dale Griffin, and Daniel Kahneman (Cambridge: Cambridge University Press, 2002), 49–81, https://doi.org/10.1017/CBO9780511808098.004.

25. Moreover, it seems highly likely that even the forced-choice recognition finding underestimated participants' "awareness." As already noted, the true trajectory of gaze when the ball is coming to the eye is increasing at a decreasing rate. Yet when Reed and colleagues actually measured gaze with a video camera, for many cases the difference between (a) α going *up at a decreasing rate* and (b) α going *up at a steady rate* was negligible, and probably led to differences in the final position of the ball that were well within the tolerance of the hand in adjusting to actually catching a ball. Despite this, a and b were two distinct alternative choice options given to participants, whose responses were scored as incorrect if they chose b rather than a. Quite apart from the possibility that participants did not attend to the subtle semantic difference between the two options, the case for classifying them as unaware on the basis that they reported that α followed a steadily increasing profile seems debatable.

26. For a critical review of several decades of research on implicit learning, see David R. Shanks, "Implicit Learning," in *Handbook of Cognition*, ed. Koen Lamberts and Robert L. Goldstone (London: Sage, 2005), 202–220.

27. For an experiment adopting essentially this method, see Darrin O. Wijeyaratnam, Zacharie Cheng-Boivin, Richard D. Bishouty, and Erin K. Cressman, "The Influence of

Awareness on Implicit Visuomotor Adaptation," *Consciousness and Cognition* 99 (2022): 103297, https://doi.org/10.1016/j.concog.2022.103297.

28. Several examples of this cycle of evidence and counterevidence are described in Shanks, "Implicit learning."

29. See Melvyn A. Goodale and A. David Milner, "Separate Visual Pathways for Perception and Action," *Trends in Neurosciences* 15, no. 1 (1992): 20–25, https://doi.org/10.1016/0166-2236(92)90344-8; M. A. Goodale, A. D. Milner, L. S. Jakobson, and D. P. Carey, "A Neurological Dissociation between Perceiving Objects and Grasping Them," *Nature* 349 (1991): 154–156, https://doi.org/10.1038/349154a0.

30. This evidence is reviewed in Thomas Schenk, "An Allocentric Rather Than Perceptual Deficit in Patient D.F.," *Nature Neuroscience* 9 (2006): 1369–1370, https://doi.org/10.1038/nn1784; Thomas Schenk and Robert D. McIntosh, "Do We Have Independent Visual Streams for Perception and Action?" *Cognitive Neuroscience* 1, no. 1 (2010): 52–62, https://doi.org/10.1080/17588920903388950.

31. See Volker H. Franz and Karl R. Gegenfurtner, "Grasping Visual Illusions: Consistent Data and No Dissociation," *Cognitive Neuropsychology* 25, nos. 7–8 (2008): 920–950, https://doi.org/10.1080/02643290701862449; Karl K. Kopiske, Nicola Bruno, Constanze Hesse, Thomas Schenk, and Volker H. Franz, "The Functional Subdivision of the Visual Brain: Is There a Real Illusion Effect on Action? A Multi-Lab Replication Study," *Cortex* 79 (2016): 130–152, https://doi.org/doi.org/10.1016/j.cortex.2016.03.020.

32. Salvatore Aglioti, Joseph F. X. Desouza, and Melvyn A. Goodale, "Size-Contrast Illusions Deceive the Eye But Not the Hand," *Current Biology* 5, no. 6 (1995): 679–685, https://doi.org/10.1016/S0960-9822(95)00133-3.

33. V. H. Franz, K. R. Gegenfurtner, H. H. Bülthoff, and M. Fahle, "Grasping Visual Illusions: No Evidence for a Dissociation between Perception and Action," *Psychological Science* 11, no. 1 (2000): 20–25, https://doi.org/10.1111/1467-9280.00209.

34. David A. Westwood and Melvyn A. Goodale, "Converging Evidence for Diverging Pathways: Neuropsychology and Psychophysics Tell the Same Story," *Vision Research* 51, no. 8 (2011): 804–811, https://doi.org/10.1016/j.visres.2010.10.014; Thomas Schenk, Volker Franz, and Nicola Bruno, "Vision-for-Perception and Vision-for-Action: Which Model Is Compatible with the Available Psychophysical and Neuropsychological Data?" *Vision Research* 51, no. 8 (2011): 812–818. https://doi.org/10.1016/j.visres.2011.02.003; Kopiske et al., "The Functional Subdivision."

35. See Lawrence Weiskrantz, *Blindsight: A Case Study and Implications* (Oxford: Oxford University Press, 1986).

36. Important statements of and responses to this view can be found in John R. Campion, Richard Latto, and Y. M. Smith, "Is Blindsight an Effect of Scattered Light, Spared Cortex, and Near-Threshold Vision?" *Behavioral and Brain Sciences* 6, no. 3

(1983): 423–486, https://doi.org/10.1017/S0140525X00016861; Larry Weiskrantz, "Is Blindsight Just Degraded Normal Vision?" *Experimental Brain Research* 192 (2009): 413–416, https://doi.org/10.1007/s00221-008-1388-7; Ian Phillips, "Blindsight Is Qualitatively Degraded Conscious Vision," *Psychological Review* 128, no. 3 (2021): 558–584, https://doi.org/10.1037/rev0000254.

37. Alan Cowey, "The Blindsight Saga," *Experimental Brain Research* 200 (2010): 7, https://doi.org/10.1007/s00221-009-1914-2.

38. For an entertaining history of the Vicary story, see Anthony R. Pratkanis, "The Cargo-Cult Science of Subliminal Persuasion," *Skeptical Inquirer* 16, no. 3 (1992): 260–272. The Judas Priest trial is described in Timothy E. Moore, "Scientific Consensus and Expert Testimony: Lessons from the Judas Priest Trial," *Skeptical Inquirer* 20 (1996): 32–38.

39. For a review of early research on this topic, see John R. Vokey and J. Don Read, "Subliminal Messages: Between the Devil and the Media," *American Psychologist* 40 (1985): 1231–1239, https://doi.org/10.1037/0003-066X.40.11.1231.

40. The work of Morten Overgaard of Aarhus University and his colleagues has been particularly convincing about this crucial point. See Michael Lohse and Morten Overgaard, "Emotional Priming Depends on the Degree of Conscious Experience," *Neuropsychologia* 128 (2019): 96–102, https://doi.org/10.1016/j.neuropsychologia.2017.10.028; Morten Overgaard, "Visual Experience and Blindsight: A Methodological Review," *Experimental Brain Research* 209 (2011): 473–479, https://doi.org/10.1007/s00221-011-2578-2; Morten Overgaard, Katrin Fehl, Kim Mouridsen, Bo Bergholt, and Axel Cleeremans "Seeing without Seeing? Degraded Conscious Vision in a Blindsight Patient," *PLOS ONE* 3, no. 8 (2008): e3028, https://doi.org/10.1371/journal.pone.0003028; Thomas Z. Ramsøy and Morten Overgaard, "Introspection and Subliminal Perception," *Phenomenology and the Cognitive Sciences* 3 (2004): 1–23, https://doi.org/10.1023/B:PHEN.0000041900.30172.e8. For further discussions on relevant methodological issues in research on subliminal perception, see David R. Shanks, Simone Malejka, and Miguel A. Vadillo, "The Challenge of Inferring Unconscious Mental Processes," *Experimental Psychology* 68, no. 3 (2021): 113–129, https://doi.org/10.1027/1618-3169/a000517; Thomas Schmidt, "Invisible Stimuli, Implicit Thresholds: Why Invisibility Judgments Cannot Be Interpreted in Isolation," *Advances in Cognitive Psychology* 11, no. 2 (2015): 31–41, https://doi.org/10.5709/acp-0169-3; Hagar Gelbard-Sagiv, Nathan Faivre, Liad Mudrik, and Christof Koch, "Low-Level Awareness Accompanies 'Unconscious' High-Level Processing during Continuous Flash Suppression," *Journal of Vision* 16, no. 1 (2016): 3, https://doi.org/10.1167/16.1.3.

41. For example, see K. Anders Ericsson and Herbert A. Simon, *Protocol Analysis: Verbal Reports as Data* (Cambridge, MA: MIT Press, 1984); Ben R. Newell and David R. Shanks, "Unconscious Influences on Decision Making: A Critical Review," *Behavioral and Brain Sciences* 37, no. 1 (2014): 1–19, https://doi.org/10.1017/s0140525x12003214.

Chapter 3

1. This seminal demonstration is reported in Tory E. Higgins, William S. Rholes, and Carl R. Jones, "Category Accessibility and Impression Formation," *Journal of Experimental Social Psychology* 13, no. 2 (1977): 141–154, https://doi.org/10.1016/s0022-1031(77)80007-3. Higgins and colleagues describe some earlier related research that set the scene for their pioneering study.

2. This follow-up research, employing slightly different procedures but just as influential as the study by Higgins et al., is Thomas K. Srull and Robert S. Wyer, "The Role of Category Accessibility in the Interpretation of Information about Persons: Some Determinants and Implications," *Journal of Personality and Social Psychology* 37, no. 10 (1979): 1660–1672, https://doi.org/10.1037/0022-3514.37.10.1660. It is hard to exaggerate the influence of these two studies. Each has been cited (on Google Scholar as of January 2022) around twenty-five hundred times, meaning that this number of later research articles have cited—and hence presumably been influenced by—the Higgins et al. and Srull and Wyer studies. We revisit these two studies later in the chapter.

3. Daniel Kahneman, *Thinking, Fast and Slow* (New York: Farrar, Straus and Giroux, 2011), 57.

4. Semantic priming was first demonstrated by David E. Meyer and Roger W. Schvaneveldt, "Facilitation in Recognizing Pairs of Words: Evidence of a Dependence between Retrieval Operations," *Journal of Experimental Psychology* 90, no. 2 (1971): 227–234, https://doi.org/10.1037/h0031564.

5. Robert J. Sternberg and Karin Sternberg, *Cognitive Psychology*, 6th ed. (Belmont, MA: Wadsworth/Cengage Learning, 2012).

6. Richard L. Gregory, "The Medawar Lecture 2001 Knowledge for Vision: Vision for Knowledge," *Philosophical Transactions of the Royal Society B: Biological Sciences* 360, no. 1458 (2005): 1231–1251, https://doi.org/10.1098/rstb.2005.1662.

7. James H. Neely, "Priming," in *Encyclopedia of Cognitive Science*, ed. Lynn Nadel (London: Nature Publishing Group, 2003), 721.

8. Kahneman, *Thinking, Fast and Slow*, 53.

9. Chris Loersch and B. Keith Payne, "The Situated Inference Model," *Perspectives on Psychological Science* 6, no. 3 (2011): 234–252, https://doi.org/10.1177/1745691611406921.

10. The original papers citing these effects are "professors": Ap Dijksterhuis and Ad van Knippenberg, "The Relation between Perception and Behavior, or How to Win a Game of Trivial Pursuit," *Journal of Personality and Social Psychology* 74, no. 4 (1998): 865–877, https://doi.org/10.1037/0022-3514.74.4.865; "graphing" and "warmth": Lawrence E. Williams & John A. Bargh, "Keeping One's Distance: The Influence of Spatial Distance Cues on Affect and Evaluation," *Psychological Science* 19, no. 3 (2008): 302–308, https://doi.org/10.1111/j.1467-9280.2008.02084.x. Lawrence E. Williams

and John A. Bargh, "Experiencing Physical Warmth Promotes Interpersonal Warmth," *Science* 322, no. 5901 (2008): 606–607, https://doi.org/10.1126/science.1162548 . Follow-up work questioning the reliability of the findings includes Harold Pashler, Noriko Coburn, and Christine R. Harris, "Priming of Social Distance? Failure to Replicate Effects on Social and Food Judgments," *PLOS ONE* 7, no. 8 (2012), https://doi.org/10.1371/journal.pone.0042510, David R. Shanks et al., "Priming Intelligent Behavior: An Elusive Phenomenon," *PLOS ONE* 8, no. 4 (2013), https://doi.org/10.1371/journal.pone.0056515; and Christopher F. Chabris, Patrick R. Heck, Jaclyn Mandart, Daniel J. Benjamin, and Daniel J. Simons, "No Evidence That Experiencing Physical Warmth Promotes Interpersonal Warmth," *Social Psychology* 50, no. 2 (2019): 127–132, https://doi.org/10.1027/1864-9335/a000361.

11. Various discussions, perspectives, and reviews can be found in Scott O. Lilienfeld and Irwin D. Waldman, *Psychological Science under Scrutiny: Recent Challenges and Proposed Solutions* (Chichester: Wiley, 2017).

12. The walking study, one of the most controversial experiments in behavioral science (cited over six thousand times), is John A. Bargh, Mark Chen, and Lara Burrows, "Automaticity of Social Behavior: Direct Effects of Trait Construct and Stereotype Activation on Action," *Journal of Personality and Social Psychology* 71, no. 2 (1996): 230–244, https://doi.org/10.1037/0022-3514.71.2.230.

13. Two large-scale studies that demonstrate this key point are Tal Moran et al., "Incidental Attitude Formation via the Surveillance Task: A Preregistered Replication of the Olson and Fazio (2001) Study," *Psychological Science* 32 (2021): 120–131, https://doi.org/10.1177/0956797620968526; and Sean Hughes, Jamie Cummins, and Ian Hussey, "Effects on the Affect Misattribution Procedure Are Strongly Moderated by Awareness," PsyArXiv, March 18, 2021, https://psyarxiv.com/d5zn8/.

14. Stéphane Doyen, Olivier Klein, Cora-Lise Pichon, and Axel Cleeremans, "Behavioral Priming: It's All in the Mind, But Whose Mind?" *PLOS ONE* 7, no. 1 (2012), https://doi.org/10.1371/journal.pone.0029081.

15. There do not appear to have been any direct, large-sample replications of the Higgins et al. experiments. However, as noted in note 2 in chapter 3, Srull and Wyer reported a variation on the Donald procedure, and this influential study has been the subject of two major large-sample replication efforts, neither of which obtained a meaningful priming effect: Randy J. McCarthy et al., "Registered Replication Report on Srull and Wyer (1979)," *Advances in Methods and Practices in Psychological Science* 1 (2018): 321–336, https://doi.org/10.1177/2515245918777487; Randy McCarthy et al. "A Multi-Site Collaborative Study of the Hostile Priming Effect," *Collabra: Psychology* 7, no. 1 (2021), https://doi.org/10.1525/collabra.18738.

16. For another striking example of how priming effects can be fully explained by experimental demands, see Thandiwe S. E. Gilder and Erin A. Heerey, "The Role of Experimenter Belief in Social Priming," *Psychological Science* 29 (2018): 403–417, https://doi.org/10.1177/0956797617737128.

17. For additional details on methods for assessing awareness, see Ben R. Newell and David R. Shanks, "Unconscious Influences on Decision Making: A Critical Review," *Behavioral and Brain Sciences* 37, no. 1 (2014): 1–19, https://doi.org/10.1017/s0140525x12003214.

18. The pen experiment, as highly cited as the two unrelated-studies experiments described earlier, is reported in Fritz Strack, Leonard L. Martin, and Sabine Stepper, "Inhibiting and Facilitating Conditions of the Human Smile: A Nonobtrusive Test of the Facial Feedback Hypothesis," *Journal of Personality and Social Psychology* 54, no. 5 (1988): 768–777, https://doi.org/10.1037/0022-3514.54.5.768.

19. Eric-Jan Wagenmakers et al., "Registered Replication Report: Strack, Martin, & Stepper (1988)," *Perspectives on Psychological Science* 11, no. 6 (2016): 917–928, https://doi.org/10.1177/1745691616674458.

20. Fritz Strack, "Reflection on the Smiling Registered Replication Report," *Perspectives on Psychological Science* 11, no. 6 (2016): 929–930, https://doi.org/10.1177/1745691616674460.

21. A more recent multilab replication attempt, that included Strack, found "inconclusive evidence" for an effect of the pen-in-the-mouth task, a conclusion that might be considered generous. See Nicholas A. Coles et al., "A Multi-Lab Test of the Facial Feedback Hypothesis by the Many Smiles Collaboration," PsyArXiv, February 4, 2019, https://doi:10.31234/osf.io/cvpuw.

22. Ulrich Schimmack, "Reconstruction of a Train Wreck: How Priming Research Went off the Rails," Replicability-Index, December 31, 2020, https://replicationindex.com/2017/02/02/reconstruction-of-a-train-wreck-how-priming-research-went-of-the-rails/.

23. Kahneman's comments can be found in Schimmack, "Reconstruction of a Train Wreck."

24. Discussion of the ongoing fallout from Kahneman's discussion of the replicability of certain priming studies can be found in the following article and its associated commentaries: Jeffrey W. Sherman and Andrew M. Rivers, "There's Nothing Social about Social Priming: Derailing the 'Train Wreck,'" *Psychological Inquiry* 32, no. 1 (2021): 1–11, https://doi.org/10.1080/1047840x.2021.1889312. We also revisit this critique in chapter 7.

25. David R. Shanks et al., "Romance, Risk, and Replication: Can Consumer Choices and Risk-Taking Be Primed by Mating Motives?" *Journal of Experimental Psychology: General* 144, no. 6 (2015), https://doi.org/10.1037/xge0000116.

Chapter 4

1. This chapter draws on several articles by Craig McKenzie and Shlomi Sher. Key references are Shlomi Sher and Craig R. M. McKenzie, "Information Leakage from Logically Equivalent Frames," *Cognition 101*, no. 3 (2006): 467–494, https://doi.org

/10.1016/j.cognition.2005.11.001; Shlomi Sher and Craig R. M. McKenzie, "Framing Effects and Rationality," in *The Probabilistic Mind: Prospects for a Bayesian Cognitive Science*, ed. Nick Chater and Mike Oaksford (Oxford: Oxford University Press, 2009), 79; Craig R. M. McKenzie, Shlomi Sher, Lin M. Leong, and Johannes Müller-Trede, "Constructed Preferences, Rationality, and Choice Architecture," *Review of Behavioral Economics* 5, no. 3–4 (2018): 337–370, https://doi.org/10.1561/105.00000091.

2. Like many of the most eye-catching peculiarities about human judgment and decision making, framing was discovered by Amos Tversky and Daniel Kahneman: see Amos Tversky and Daniel Kahneman, "The Framing of Decisions and the Psychology of Choice," *Science* 211 (1981): 453–458, https://doi.org/10.1126/science.7455683.

3. An interesting aside here is that in 2011, the US government mandated that companies could only claim products to be 90 percent fat free if they also said they were 10 percent fat! For further discussion, see Cass R Sunstein, "Nudges That Fail," *Behavioral Public Policy* 1, no. 1 (2017): 4–25, https://doi.org/10.1017/bpp.2016.3.

4. A comprehensive discussion of how framing effects operate can be found in Irwin P. Levin, Sandra L. Schneider, and Gary J. Gaeth, "All Frames Are Not Created Equal: A Typology and Critical Analysis of Framing Effects," *Organizational Behavior and Human Decision Processes* 76, no. 2 (1998): 149–188, https://doi.org/10.1006/obhd.1998.2804.

5. The paper that introduced framing effects to the medical literature is Barbara J. McNeil, Stephen G. Pauker, Harold C. Sox Jr., and Amos Tversky, "On the Elicitation of Preferences for Alternative Therapies," *New England Journal of Medicine* 306, no. 21 (1982): 1259–1262, https://doi.org/10.1056/NEJM198205273062103.

6. Sher and McKenzie, "Information Leakage."

7. Sher and McKenzie, "Information Leakage."

8. Sher and McKenzie, "Information Leakage."

9. For further discussion of what might determine the salience of different attributes in people's psycholinguistic representations and more detail on the *reference-point hypothesis*, see Sher and McKenzie, "Framing Effects and Rationality."

10. Many studies have asked whether the position of an item on a menu affects diners' choices, and such an effect (if real) could be exploited as a useful nudge for encouraging healthy eating. In fact, the evidence is rather mixed, with some studies finding effects and others not. For two examples, one positive and the other not, see Eran Dayan and Maya Bar-Hillel, "Nudge to Nobesity II: Menu Positions Influence Food Orders," *Judgment and Decision Making* 6, no. 4 (2011): 333–342; and Rebecca Wyse et al., "Can Changing the Position of Online Menu Items Increase Selection of Fruit and Vegetable Snacks? A Cluster Randomized Trial within an Online Canteen Ordering System in Australian Primary Schools," *American Journal of Clinical Nutrition* 109, no. 5 (2019): 1422–1430, https://doi.org/10.1093/ajcn/nqy351.

11. Sher and McKenzie, "Information Leakage," 489.

12. This experiment is reported in Balazs Aczel, Aba Szollosi, and Bence Bago, "The Effect of Transparency on Framing Effects in within-Subject Designs," *Journal of Behavioral Decision Making* 31, no. 1 (2018): 25–39, https://doi.org/10.1002/bdm.2036.

13. The terms *choice architecture* and *nudge* were popularized in the hugely influential book by Richard Thaler and Cass R. Sunstein, *Nudge: Improving Decisions about Health, Wealth, and Happiness* (New Haven, CT: Yale University Press, 2008). A nudge is defined as any aspect of the choice architecture that alters people's behavior in a predictable way without forbidding any options or significantly changing their economic incentives.

14. See Eric J. Johnson and Daniel Goldstein, "Do Defaults Save Lives?" *Science* 302, no. 5649 (2003): 1338–1339, https://doi.org/10.1126/science.1091721. This has become one of the most famous examples of how simple changes to a choice architecture (the setting of the default) can have a large impact on organ donation. For a more contemporary discussion of this and other influences of choice architecture, see Eric J. Johnson, *The Elements of Choice* (New York: Riverhead Books, 2021).

15. For the original versions and more details of these survey questions, see Craig R. M. McKenzie, Michael J. Liersch, and Stacey R. Finkelstein, "Recommendations Implicit in Policy Defaults," *Psychological Science* 17, no. 5 (2006): 414–420, https://doi.org/10.1111/j.1467-9280.2006.01721.x.

16. Although the idea that people can interpret the intentions behind default settings seems relatively uncontroversial, some work has questioned whether people know how to set defaults in order to influence particular outcomes. See Julian J. Zlatev, David P. Daniels, Hajin Kim, and Margaret A. Neale, "Default Neglect in Attempts at Social Influence," *Proceedings of the National Academy of Sciences* 114, no. 52 (2017): 13643–48, https://doi.org/10.1073/pnas.1712757114; and the comment by Minah Jung and colleagues that challenges the generality of the claims made by Zlatev et al.: Minah H. Jung, Chengyao Sun, and Leif D. Nelson, "People Can Recognize, Learn, and Apply Default Effects in Social Influence," *Proceedings of the National Academy of Sciences* 115, no. 35 (2018): E8105–E8106, https://doi.org/10.1073/pnas.1810986115.

17. This example is adapted from experiments reported in Craig R. M. McKenzie, Lim M. Leong, and Shlomi Sher, "Default Sensitivity in Attempts at Social Influence," *Psychonomic Bulletin & Review* 28, no 2 (2020): 695–702, https://doi.org/10.3758/s13423-020-01834-4.

18. This text comes from an experiment reported by Yavor Paunov, Michaela Wänke, and Tobias Vogel, "Ethical Defaults: Which Transparency Components Can Increase the Effectiveness of Default Nudges?" *Social Influence* 14, no. 3–4 (2019): 104–116, https://doi.org/10.1080/15534510.2019.1675755.

19. Discussions about the (in)effectiveness of defaults and nudging more generally as a tool for behavioral change go beyond what we can consider here. For illuminating perspectives on people's "nudgeability," see Denise de Ridder, Floor Kroese, and

Laurens van Gestel, "Nudgeability: Mapping Conditions of Susceptibility to Nudge Influence," *Perspectives on Psychological Science* 17, no. 2 (2021): 346–359, https://doi.org/10.1177/1745691621995183; and Job M. T. Krijnen, David Tannenbaum, and Craig R. Fox, "Choice Architecture 2.0: Behavioral Policy as an Implicit Social Interaction," *Behavioral Science and Policy* 3, no. 2 (2017): i–18, https://doi.org/10.1353/bsp.2017.0010.

20. Amos Tversky and Daniel Kahneman, "Judgment under Uncertainty: Heuristics and Biases," *Science* 185, no. 4157 (1974): 1124–1131, https://doi.org/10.1126/science.185.4157.1124.

21. Denis J. Hilton, "The Social Context of Reasoning: Conversational Inference and Rational Judgment," *Psychological Bulletin* 118, no. 2 (1995): 248–271, https://doi.org/10.1037/0033-2909.118.2.248.

22. Hilton, "Social Context of Reasoning."

23. Joseph P. Simmons, Robyn A. LeBoeuf, and Leif D. Nelson, "The Effect of Accuracy Motivation on Anchoring and Adjustment: Do People Adjust from Provided Anchors?" *Journal of Personality and Social Psychology* 99, no. 6 (2010): 917–932, https://doi.org/10.1037/a0021540.

24. A review of the effects of uninformative anchors on payments for consumer goods is provided by Konstantinos Ioannidis, Theo Offerman, and Randolph Sloof, "On the Effect of Anchoring on Valuations When the Anchor Is Transparently Uninformative," *Journal of the Economic Science Association* 6, no. 1 (2020): 77–94, https://doi.org/10.1007/s40881-020-00094-1, who in fact present meta-analytic evidence that anchoring effects are completely absent when people truly believe that the anchor value is random. On anchor plausibility, see Andrew R. Smith, Paul D. Windschitl, and Kathryn Bruchmann, "Knowledge Matters: Anchoring Effects Are Moderated by Knowledge Level," *European Journal of Social Psychology* 43 (2013): 97–108, https://doi.org/10.1002/ejsp.1921.

25. For evidence regarding people's awareness of their use of the anchor and supporting the claim that anchors only affect judgments when they are regarded as providing good estimates, see Adam J. L. Harris, Tom Phillips, Sam Bhaskaran, Jozefina Krasniqi, and David R. Shanks, "Awareness of Anchoring," unpublished report (2022); and Oliver Schweickart, Cory Tam, and Norman R. Brown, "When "Bad" Is Good: How Evaluative Judgments Eliminate the Standard Anchoring Effect," *Canadian Journal of Experimental Psychology* 75, no. 1 (2021): 56–63, https://doi.org/10.1037/cep0000209.

26. See Sangsuk Yoon, Nathan M. Fong, and Angelika Dimoka, "The Robustness of Anchoring Effects on Preferential Judgments," *Judgment and Decision Making* 14, no. 4 (2019), for a comprehensive analysis of the impact of random anchors (among other factors) and also Minah H. Jung, Hannah Perfecto, and Leif D. Nelson, "Anchoring in Payment: Evaluating a Judgmental Heuristic in Field Experimental Settings," *Journal of Marketing Research* 53, no. 3 (2016): 354–368, https://doi.org/10.1509/jmr.14

.0238, for additional relevant findings. One of our own articles examines the specific role of incidental anchors and finds the evidence lacking: David R. Shanks, Pietro Barbieri-Hermitte, and Miguel A. Vadillo, "Do Incidental Environmental Anchors Bias Consumers' Price Estimations?" *Collabra: Psychology* 6, no. 1 (2020): 19, https://doi.org/10.1525/collabra.310.

27. This thought experiment is described by the well-known judgment and decision-making scholar Robin Hogarth in a comment on an article we wrote in 2014: Ben R. Newell and David R. Shanks, "Unconscious Influences on Decision Making: A Critical Review," *Behavioral and Brain Sciences* 37, no. 1 (2014): 1–19, https://doi.org/10.1017/s0140525x12003214. He in turn attributed it to the famous social psychologist Richard Nisbett, who used the scenario in a talk Hogarth attended. In the comment, Hogarth relates that though he likes and remembers the scenario, he does not recall the specific point that Nisbett was illustrating. See Robin M. Hogarth, "Automatic Processes, Emotions, and the Causal Field," *Behavioral and Brain Sciences* 37, no. 1 (2014): 31–32, https://doi.org/10.1017/S0140525X13000757.

28. For further development of this argument see David R. Shanks and Ben R. Newell, "Authors' Response: The Primacy of Conscious Decision Making," *Behavioral and Brain Sciences* 37, no. 1 (2014): 45–61, https://doi.org/10.1017/s0140525x13001507.

29. For additional discussion of the subtleties of information leakage and some evidence questioning people's sensitivity to frames, see Adam J. L. Harris, Sarah C. Jenkins, Gloria W. S. Ma, and Aloysius Oh, "Testing the Adaptability of People's Use of Attribute Frame Information," *Cognition* 212 (2021): 104720, https://doi.org/10.1016/j.cognition.2021.104720.

Chapter 5

1. "The Female CEOs on This Year's Fortune 500 Just Broke Three All-Time Records," *Fortune*, June 2, 2021, https://fortune.com/2021/06/02/female-ceos-fortune-500-2021-women-ceo-list-roz-brewer-walgreens-karen-lynch-cvs-thasunda-brown-duckett-tiaa/.

2. "Corrective Services, Australia," Australian Bureau of Statistics, accessed December 21, 2021, https://www.abs.gov.au/statistics/people/crime-and-justice/corrective-services-australia/latest-release.

3. The relevant definition in the Oxford English Dictionary is: "Tendency to favour or dislike a person or thing, especially as a result of a preconceived opinion; partiality, prejudice . . . any preference or attitude that affects outlook or behavior, esp. by inhibiting impartial consideration or judgement."

4. The data in this section come from the following sources: "Annual Admissions Statistical Report," University of Oxford, accessed December 21, 2021, https://www.ox.ac.uk/sites/files/oxford/Annual%20Admissions%20Statistical%20Report%202020.pdf; "2019 Entry UCAS Undergraduate Reports by Sex, Area Background, and

Ethnic Group," UCAS, accessed December 21, 2021, https://www.ucas.com/file/309981/download?token=u3STriXV; "Black and Ethnic Minority Students at the University of Oxford," Full Fact, accessed December 21, 2021, https://fullfact.org/education/bme-students-oxford/.

5. "Understanding Unconscious Bias," Royal Society, accessed December 21, 2021, https://royalsociety.org/topics-policy/publications/2015/unconscious-bias/.

6. Reviews of research on the effectiveness of diversity training can be found in Hussain Alhejji, Thomas Garavan, Ronan Carbery, Fergal O'Brien, and David McGuire, "Diversity Training Programme Outcomes: A Systematic Review," *Human Resource Development Quarterly* 27, no. 1 (2016): 95–149, https://doi.org/10.1002/hrdq.21221, and Katerina Bezrukova, Chester S. Spell, Jamie L. Perry, and Karen A. Jehn, "A Meta-Analytical Integration of over 40 Years of Research on Diversity Training Evaluation," *Psychological Bulletin* 142, no. 11 (2016): 1227–1274, https://doi.org/10.1037/bul0000067.

7. This sort of backfire effect was shown by Lisa Legault, Jennifer N. Gutsell, and Michael Inzlicht, "Ironic Effects of Antiprejudice Messages: How Motivational Interventions Can Reduce (But Also Increase) Prejudice," *Psychological Science* 22, no. 12 (2011): 1472–1477, https://doi.org/10.1177/0956797611427918.

8. See Janine Willis and Alexander Todorov, "First Impressions: Making Up Your Mind after a 100-ms Exposure to a Face," *Psychological Science* 17, no. 7 (2006): 592–598, https://doi.org/10.1111/j.1467-9280.2006.01750.x; Y. Z. Foo, C. A. M. Sutherland, N. S. Burton, S. Nakagawa, and G. Rhodes, "Accuracy in Facial Trustworthiness Impressions: Kernel of Truth or Modern Physiognomy? A Meta-Analysis," *Personality and Social Psychology Bulletin* 48, no. 11 (2022): 1580–1596, https://doi.org/10.1177/01461672211048110.

9. This study is described in Arnaud Tognetti, Claire Berticat, Michel Raymond, and Charlotte Faurie, "Is Cooperativeness Readable in Static Facial Features? An Inter-Cultural Approach," *Evolution and Human Behavior* 34, no. 6 (2013): 427–432, https://doi.org/10.1016/j.evolhumbehav.2013.08.002.

10. This famous study was conducted by Corinne A. Moss-Racusin, John F. Dovidio, Victoria L. Brescoll, Mark J. Graham, and Jo Handelsman, "Science Faculty's Subtle Gender Biases Favor Male Students," *Proceedings of the National Academy of Sciences* 109, no. 41 (2012): 16474–16479, https://doi.org/10.1073/pnas.1211286109.

11. This and other important issues on gender bias in academia are reviewed in an authoritative recent review by Stephen J. Ceci, Shulamit Kahn, and Wendy M. Williams, "Some Evidence of Gender Bias in Two of Six Domains in Academic Science," *Psychological Science in the Public Interest* (2023).

12. Wendy M. Williams and Stephen J. Ceci, "National Hiring Experiments Reveal 2:1 Faculty Preference for Women on STEM Tenure Track," *Proceedings of the National Academy of Sciences* 112, no. 17 (2015): 5360–5365, https://doi.org/10.1073/pnas.1418878112.

13. Ceci, Kahn, and Williams, "Some Evidence of Gender Bias."

14. In one recent study of over 250,000 real hiring decisions in an online labor market, no evidence of a male bias was found: Jason Chan and Jing Wang, "Hiring Preferences in Online Labor Markets: Evidence of a Female Hiring Bias," *Management Science* 64, no. 7 (2018): 2973–2994, https://doi.org/10.1287/mnsc.2017.2756.

15. Ceci, Kahn, and Williams, "Some Evidence of Gender Bias."

16. This viewpoint is elaborated in Lee Jussim, Jarret T. Crawford, Stephanie M. Anglin, John R. Chambers, Sean T. Stevens, and Florette Cohen, "Stereotype Accuracy: One of the Largest Relationships and Most Replicable Effects in All of Social Psychology," in *Handbook of Prejudice, Stereotyping, and Discrimination*, 2nd ed., ed. Todd D. Nelson (Hillsdale, NJ: Erlbaum, 2016), 31–63. A robust social constructionist critique by John Dixon argues that stereotyping and prejudice emerge from institutionalized power relationships and hence become accepted ways of perceiving and treating others: John Dixon, "'Thinking Ill of Others without Sufficient Warrant?' Transcending the Accuracy-Inaccuracy Dualism in Prejudice and Stereotyping Research," *British Journal of Social Psychology* 56, no. 1 (2017): 4–27, https://doi.org/10.1111/bjso.12181.

17. Brian A. Nosek et al., "Pervasiveness and Correlates of Implicit Attitudes and Stereotypes," *European Review of Social Psychology* 18, no. 1 (2007): 36–88, https://doi.org/10.1080/10463280701489053.

18. Few other topics have proven more controversial: see Scott O. Lilienfeld, "Microaggressions: Strong Claims, Inadequate Evidence," *Perspectives on Psychological Science* 12, no. 1 (2017): 138–169, https://doi.org/10.1177/1745691616659391.

19. The available correlation evidence is reviewed by Ulrich Schimmack, "The Implicit Association Test: A Method in Search of a Construct," *Perspectives on Psychological Science* 16, no. 2 (2021): 396–414, https://doi.org/10.1177/1745691619863798. Schimmack shows, however, that when the psychometric properties of the IAT are taken into account, there is a very high correlation between the latent factors that underlie explicit and implicit attitudes. From this, he concludes that to the extent that the IAT measures anything meaningful, it measures the same thing as explicit attitude scales.

20. Gregory Mitchell and Philip E. Tetlock, "Popularity as a Poor Proxy for Utility: The Case of Implicit Prejudice," in *Psychological Science under Scrutiny: Recent Challenges and Proposed Solutions*, ed. Scott O. Lilienfeld and Irwin D. Waldman (Chichester: Wiley, 2017), 164–195.

21. Adam Hahn, Charles M. Judd, Holen K. Hirsch, and Irene V. Blair, "Awareness of Implicit Attitudes," *Journal of Experimental Psychology: General* 143, no. 3 (2014): 1369–1392, https://doi.org/10.1037/a0035028.

22. See Jan De Houwer, Tom Beckers, and Agnes Moors, "Novel Attitudes Can Be Faked on the Implicit Association Test," *Journal of Experimental Social Psychology* 43, no. 6 (2007): 972–978, https://doi.org/10.1016/j.jesp.2006.10.007; Klaus Fiedler and

Matthias Bluemke, "Faking the IAT: Aided and Unaided Response Control on the Implicit Association Tests," *Basic and Applied Social Psychology* 27, no. 4 (2005): 307–316, https://doi.org/10.1207/s15324834basp2704_3.

23. Hart Blanton, James Jaccard, Erin Strauts, Gregory Mitchell, and Philip E. Tetlock, "Toward a Meaningful Metric of Implicit Prejudice," *Journal of Applied Psychology* 100, no. 5 (2015): 1468–1481, https://doi.org/10.1037/a0038379.

24. Mitchell and Tetlock, "Popularity as a Poor Proxy."

25. Anthony G. Greenwald and Calvin K. Lai, "Implicit Social Cognition," *Annual Review of Psychology* 71 (2020): 419–445, https://doi.org/10.1146/annurev-psych-010419-050837. A perfectly reliable test would give exactly the same answer on two occasions. Greenwald and Lai estimated the test-retest reliability of the IAT at 0.50. What does this number mean in practice? A value of 0.50 can be interpreted in the following way. Imagine that two people, A and B, take an IAT and the test yields a higher score for A than B. If they take the test again, the probability that A will again score higher than B is a very modest 67 percent, against a chance score of 50 percent if the accuracy of the test is no better than tossing a coin.

26. Yoav Bar-Anan and Michelangelo Vianello, "A Multimethod Multi-Trait Test of the Dual-Attitude Perspective," *Journal of Experimental Psychology: General* 147, no. 8 (2018): 1264–1272, https://doi.org/10.1037/xge0000383.

27. Klaus Rothermund and Dirk Wentura, "Underlying Processes in the Implicit Association Test: Dissociating Salience from Associations," *Journal of Experimental Psychology: General* 133, no. 2 (2004): 139–165, https://doi.org/10.1037/0096-3445.133.2.139.

28. Rickard Carlsson and Jens Agerström, "A Closer Look at the Discrimination Outcomes in the IAT Literature," *Scandinavian Journal of Psychology* 57, no. 4 (2016): 278–287, https://doi.org/10.1111/sjop.12288.

29. See Hart Blanton, Christopher N. Burrows, and James Jaccard, "To Accurately Estimate Implicit Influences on Health Behavior, Accurately Estimate Explicit Influences," *Health Psychology* 35, no. 8 (2016): 856–860, https://doi.org/10.1037/hea0000348, and Frederick L. Oswald, Gregory Mitchell, Hart Blanton, James Jaccard, and Philip E. Tetlock, "Predicting Ethnic and Racial Discrimination: A Meta-Analysis of IAT Criterion Studies," *Journal of Personality and Social Psychology* 105, no. 2 (2013): 171–192, https://doi.org/10.1037/a0032734.

30. Patrick S. Forscher et al., "A Meta-Analysis of Procedures to Change Implicit Measures," *Journal of Personality and Social Psychology* 117, no. 3 (2019): 522–559, https://doi.org/10.1037/pspa0000160.

31. The studies are Teresa J. Rosegrant and James C. McCroskey, "The Effects of Race and Sex on Proxemic Behavior in an Interview Setting," *Southern Journal of Communication* 40, no. 4 (1975): 408–418, https://doi.org/10.1080/10417947509372282; and Carl O. Word, Mark P. Zanna, and Joel Cooper, "The Nonverbal Mediation of

Self-Fulfilling Prophecies in Interracial Interaction," *Journal of Experimental Social Psychology* 10, no. 2 (1974): 109–120, https://doi.org/10.1016/0022-1031(74)90059-6.

32. "Understanding Unconscious Bias," Royal Society, accessed December 21, 2021, https://royalsociety.org/topics-policy/publications/2015/unconscious-bias/.

33. Rindy C. Anderson and Casey A. Klofstad, "Preference for Leaders with Masculine Voices Holds in the Case of Feminine Leadership Roles," *PLOS ONE* 7, no. 12 (2012): e51216, https://doi.org/10.1371/journal.pone.0051216.

Chapter 6

1. This quotation from Franklin's writings comes from University of Virginia Press, "Founders Online: From Benjamin Franklin to Joseph Priestley, 19 September 1772," accessed January 27, 2022, https://founders.archives.gov/documents/Franklin/01-19-02-0200. The original source is *The Papers of Benjamin Franklin*, vol. 19, *January 1 through December 31, 1772*, ed. William B. Willcox (New Haven: Yale University Press, 1975).

2. This page from Darwin's journal is part of a treasure trove of information at the Darwin Correspondence Project, "Darwin on Marriage," accessed January 27, 2022, https://www.darwinproject.ac.uk/tags/about-darwin/family-life/darwin-marriage.

3. For a comprehensive discussion of contemporary formal and prescriptive approaches to decision making, see Ben R. Newell, David A. Lagnado, and David R. Shanks, *Straight Choices: The Psychology of Decision Making*, 3rd ed. (London: Psychology Press, 2022).

4. This is an excerpt from a sketch by the British comedians, actors, and writers Stephen Fry and Hugh Laurie. In the sketch, Laurie is a patient (Mr. Pepperdyne) visiting a doctor (played by Fry) who tries to convince Laurie to take a course of cigarettes to help with his breathing. The excerpt is Fry's response to Laurie's query about whether too much cholesterol is bad for you. Stephen Fry and Hugh Laurie, "Doctor Tobacco," A Bit of Fry and Laurie," accessed January 28, 2022, https://abitoffryandlaurie.co.uk/sketches/doctor_tobacco.

5. The magic number 7 (plus or minus 2) is a reference to one of the most influential articles in the psychology of memory: George A. Miller, "The Magical Number Seven Plus or Minus Two: Some Limits on Our Capacity for Processing Information," *Psychological Review* 63, no. 2 (1956): 81–97, https://doi.org/10.1037/h0043158.

6. The jam study, along with other demonstrations of the purported dangers of introspection, is published in Timothy D. Wilson and Jonathan W. Schooler, "Thinking Too Much: Introspection Can Reduce the Quality of Preferences and Decisions," *Journal of Personality and Social Psychology* 60, no. 2 (1991): 181–192, https://doi.org/10.1037//0022-3514.60.2.181.

7. Wilson and Schooler, "Thinking Too Much."

8. Ap Dijksterhuis cites many of these anecdotal examples in support of the basic tenets of unconscious thought theory in Ap Dijksterhuis and Madelijn Strick, "A Case for Thinking without Consciousness," *Perspectives on Psychological Science* 11, no. 1 (2016): 117–132, https://doi.org/10.1177/1745691615615317.

9. David Kadavy, "'Yesterday' Came to Paul McCartney in a Dream. Was It a Creative Miracle?" *Getting Art Done,* April 16, 2018, https://medium.com/getting-art-done/yesterday-came-to-paul-mccartney-in-a-dream-was-it-a-creative-miracle-79839cb303fe.

10. Ap Dijksterhuis, Maarten W. Bos, Loran F. Nordgren, and Rick B. van Baaren, "On Making the Right Choice: The Deliberation-without-Attention Effect" *Science,* 311, no. 5763 (2006): 1005–1007, https://doi.org/10.1126/science.1121629. This article introduced the idea that "unconscious thought" could be beneficial for decision making. It sparked a good deal of controversy and was followed by several failures to replicate the apparent advantages of not thinking (see notes 11 and 12).

11. This pattern of results was found by many researchers seeking to replicate the unconscious thought effect, including one of our studies: Ben R. Newell, Kwan Yao Wong, Jeremy C. H. Cheung, and Tim Rakow, "Think, Blink or Sleep on It? The Impact of Modes of Thought on Complex Decision Making," *Quarterly Journal of Experimental Psychology* 62, no. 4 (2009): 707–732, https://doi.org/10.1080/17470210802215202.

12. Newell et al., "Think, Blink or Sleep on It?"; Wen Ying Moi and David R. Shanks, "Can Lies Be Detected Unconsciously?" *Frontiers in Psychology* 6 (2015): 1221, https://doi.org/10.3389/fpsyg.2015.01221.

13. Mark R. Nieuwenstein. Tjardie Wierenga, Richard D. Morey, Jelte M. Wicherts, Tessa N. Blom, Eric-Jan Wagenmakers, and Hedderik van Rijn, "On Making the Right Choice: A Meta-Analysis and Large-Scale Replication Attempt of the Unconscious Thought Advantage," *Judgment and Decision Making* 10, no. 1 (2015): 1–17.

14. Malcolm Gladwell, *Blink: The Power of Thinking without Thinking* (Harlow: Penguin Books, 2006). Gladwell cites Ambady's work extensively in his development of the idea that split-second, intuitive judgments can be highly accurate.

15. The original study is reported in Nalini Ambady and Robert Rosenthal, "Half a Minute: Predicting Teacher Evaluations from Thin Slices of Nonverbal Behavior and Physical Attractiveness," *Journal of Personality and Social Psychology* 64, no. 3 (1993): 431–441, https://doi.org/10.1037//0022-3514.64.3.431.

16. Gladwell, *Blink,* 23.

17. The husband-and-wife team of John and Julie Gottman run a relationship counseling business that applies the "Gottman method." For details, see the Gottman Institute (www.gottman.com).

18. Robert J. Waldinger, Marc S. Schulz, Stuart T. Hauser, Joseph P. Allen, and Judith A. Crowell, "Reading Others' Emotions: The Role of Intuitive Judgments in Predicting Marital Satisfaction, Quality, and Stability," *Journal of Family Psychology* 18, no. 1 (2004): 58–71, https://doi.org/10.1037/0893-3200.18.1.58.

19. Similar to unconscious thought theory, some researchers have questioned the validity of Gottman's whole prediction enterprise. For a quick summary, see Laurie Abraham, "A Dissection of John Gottman's Love Lab," *Slate*, March 8, 2010, https://slate.com/human-interest/2010/03/a-dissection-of-john-gottman-s-love-lab.html. For a more detailed analysis of the statistical problems that beset accurate forecasting of marriages, see Richard E. Heyman and Amy M. Smith Slep, "The Hazards of Predicting Divorce without Cross-Validation," *Journal of Marriage and the Family* 63, no. 2 (2001): 473–479, https://doi.org/10.1111/j.1741-3737.2001.00473.x.

20. Nalini Ambady, "The Perils of Pondering: Intuition and Thin Slice Judgments," *Psychological Inquiry* 21, no. 4 (2010): 271–278, https://doi.org/10.1080/1047840x.2010.524882.

21. Ambady, "The Perils of Pondering," 276.

22. The study by Ambady (2010) discussed in the previous section does appear to offer some evidence that providing reasons leads to objectively worse decisions, given that there was a ground truth regarding the relationship status of the couples. However, the sample sizes in this experiment were rather small (and the effect size implausibly large), suggesting that until replicated, it should not warrant too much attention. More research on this issue is needed. For other intriguing suggestive results about the detrimental impact of articulating reasons for a prediction, see Jamin Brett Halberstadt and Gary M. Levine, "Effects of Reasons Analysis on the Accuracy of Predicting Basketball Games," *Journal of Applied Social Psychology* 29, no. 3 (1999): 517–530, https://doi.org/10.1111/j.1559-1816.1999.tb01399.x.

23. The quotation about recognition comes from Herbert A. Simon, "What Is an 'Explanation' of Behavior?" *Psychological Science* 3, no. 3 (1992): 150–161, https://doi.org/10.1111/j.1467-9280.1992.tb00017.x. This article, based on Simon's keynote address to the annual convention of the American Psychological Society in June 1991, is a cogent and thought-provoking piece that asks fundamental questions about what it means to explain behavior.

24. In a letter to Dr. H. L. Gordon on May 3, 1949, Albert Einstein Archives 58–217, cited in Walter Isaacson, *Einstein: His Life and Universe* (New York: Simon & Schuster, 2007).

25. For more on the theory that insight is accompanied by flashes of suspicion rather than flashes of inspiration, see Nick Chater, *The Mind Is Flat: The Remarkable Shallowness of the Improvising Brain* (New Haven: Yale University Press, 2018).

26. There have been decades of research on learning and memory under anesthesia. For a recent assessment, see Victor X. Fu, Karel J. Sleurink, Joséphine C. Janssen,

Bas P. L. Wijnhoven, Johannes Jeekel, and Markus Klimek, "Perception of Auditory Stimuli during General Anesthesia and Its Effects on Patient Outcomes: A Systematic Review and Meta-Analysis," *Canadian Journal of Anesthesia* 68 (2021): 1231–1253, https://doi.org/10.1007/s12630-021-02015-0.

Chapter 7

1. This question comes from the Cognitive Reflection Test, published in Shane Frederick, "Cognitive Reflection and Decision Making," *Journal of Economic Perspectives*, 19, no. 4 (2005): 25–42, https://doi.org/10.1257/089533005775196732. The CRT, as it has become known, is an extremely popular short test for (apparently) measuring people's degree of automatic versus deliberative (reflective) processing.

2. The quotes from the World Bank Report can be found in World Bank Group, *World Development Report 2015: Mind, Society, and Behavior* (Washington, DC: World Bank, 2015), https://openknowledge.worldbank.org/handle/10986/20597.

3. See Jonathan St. B. T. Evans, "Dual-Processing Accounts of Reasoning, Judgement and Social Cognition," *Annual Review of Psychology* 59 (2008): 255–278, https://doi.org/10.1146/annurev.psych.59.103006.093629, and Steven A. Sloman, "The Empirical Case for Two Systems of Reasoning," *Psychological Bulletin* 119 (1996): 3–22, https://doi.org/10.1037/0033-2909.119.1.3.

4. See Jonathan St. B. T. Evans, and Jodie Curtis-Holmes, "Rapid Responding Increases Belief Bias: Evidence for the Dual-Process Theory of Reasoning," *Thinking and Reasoning* 11, no. 4 (2005): 382–389, https://doi.org/10.1080/13546780542000005. The discussion of Evans's experiment in this section is based on insightful work by Rachel Stephens and her colleagues: Rachel G. Stephens, Dora Matzke, and Brett K. Hayes, "Disappearing Dissociations in Experimental Psychology: Using State-Trace Analysis to Test for Multiple Processes," *Journal of Mathematical Psychology* 90 (2019): 3–22, https://doi.org/10.1016/j.jmp.2018.11.003. Much of Stephens's analysis is based on state-trace analysis, an important tool for assessing assumptions about unobservable, latent mental states.

5. These trends can be quantified. If we average the endorsement rates in the two valid conditions, valid and believable (VB) and valid and unbelievable (VU), and subtract the average endorsement rates in the two invalid conditions, invalid and believable (IB) and invalid and unbelievable (IU), we obtain a validity score of 0.33 in the no-time-pressure condition. A corresponding calculation subtracting the unbelievable (VU and IU) from the believable (VB and IB) syllogism endorsement rates yields a believability score of 0.31. In the time pressure condition, the validity score is lower, 0.13, and the believability score higher, 0.55, than their equivalents in the no time pressure condition. So time pressure reduces the impact of validity on decisions but increases the impact of believability.

6. Stephens et al., "Disappearing Dissociations in Experimental Psychology."

7. John R. Stroop, "Studies of Interference in Serial Verbal Reactions," *Journal of Experimental Psychology* 18, no. 6 (1935): 643–662, https://doi.org/10.1037/h0054651.

8. Arthur R. Jensen and William D. Rohwer Jr., "The Stroop Color-Word Test: A Review," *Acta Psychologica* 25, no. 1 (1966): 36–93, https://doi.org/10.1016/0001-6918(66)90004-7.

9. Colin M. MacLeod, "Half a Century of Research on the Stroop Effect: An Integrative Review," *Psychological Bulletin* 109, no. 2 (1991): 163–203, https://doi.org/10.1037/0033-2909.109.2.163.

10. Derek Besner and Jennifer A. Stolz, "What Kind of Attention Modulates the Stroop Effect?" *Psychonomic Bulletin & Review* 6, no. 1 (1999): 99–104, https://doi.org/10.3758/bf03210815.

11. Derek Besner and Jennifer A. Stolz, "Unconsciously Controlled Processing: The Stroop Effect Reconsidered," *Psychonomic Bulletin & Review* 6, no. 3 (1999): 449–455, https://doi.org/10.3758/bf03210834. This article presents several arguments against the idea that the Stroop effect provides evidence for automatic processing. For a recent review of this topic, see Derek Besner, "Visual Word Recognition: Attention, Intention, Context, and Processing Dynamics," *Canadian Journal of Experimental Psychology/Revue canadienne de psychologie expérimentale* 76, no. 1 (2022): 57, https://doi.org/10.1037/cep0000274.

12. Daniel Kahneman, *Thinking, Fast and Slow* (New York: Farrar, Straus and Giroux, 2011), 29.

13. For further discussion of the reality of the two systems and the dangers of dichotomies, see Olivier Corneille and Mandy Hütter, "Implicit? What Do You Mean? A Comprehensive Review of the Delusive Implicitness Construct in Attitude Research," *Personality and Social Psychology Review* 24, no. 3 (2020): 212–232, https://doi.org/10.1177/1088868320911325; Gideon Keren and Yaacov Schul, "Two Is Not Always Better Than One: A Critical Evaluation of Two-System Theories," *Perspectives on Psychological Science* 4, no. 6 (2009): 533–50. https://doi.org/10.1111/j.1745-6924.2009.01164.x; David E. Melnikoff and John A. Bargh, "The Mythical Number Two," *Trends in Cognitive Sciences* 22, no. 4 (2018): 280–293, https://doi.org/10.1016/j.tics.2018.02.001; Magda Osman, "A Case Study: Dual-Process Theories of Higher Cognition—Commentary on Evans & Stanovich," *Perspectives on Psychological Science* 8, no. 3 (2013): 248–252.

14. See Agnes Moors and Jan De Houwer, "Automaticity: A Theoretical and Conceptual Analysis," *Psychological Bulletin* 132, no. 2 (2006): 297–326, https://doi.org/10.1037/0033-2909.132.2.297.

15. Nick Chater, "Is the Type 1/Type 2 Distinction Important for Behavioral Policy?" *Trends in Cognitive Sciences* 22, no. 5 (2018): 369–371, https://doi.org/10.1016/j.tics.2018.02.007.

Chapter 8

1. Ed Yong, "Replication Studies: Bad Copy," *Nature* 485, no. 7398 (2012): 298–300, https://doi.org/10.1038/485298a.

2. A detailed assessment of the controversy surrounding these priming effects is available in the following article and its associated commentaries: Jeffrey W. Sherman and Andrew M. Rivers, "There's Nothing Social about Social Priming: Derailing the 'Train Wreck,'" *Psychological Inquiry* 32, no. 1 (2021): 1–11, https://doi.org/10.1080/1047840x.2021.1889312.

3. Diederik Stapel's memoir was originally published in Dutch in 2012 as *Ontsporing* (Amsterdam: Prometheus, 2012). The excerpts in this chapter come from an English translation by Nick Brown entitled *Faking Science: A True Story of Scientific Fraud*, accessed January 27, 2022, http://nick.brown.free.fr/stapel.

4. The full report of this study was published in the prestigious journal *Science* in 2011. It was then retracted—removed from the scientific record—later that year following the revelations about Stapel's practices. The article can be still be accessed, with the accompanying retraction notice, here: Diederik A. Stapel and Siegwart Lindenberg, "Coping with Chaos: How Disordered Contexts Promote Stereotyping and Discrimination," *Science* 332, no. 6026 (2011): 251–253, https://doi.org/10.1126/science.1201068.

5. Stapel, *Faking Science*, 119.

6. The Levelt Report was commissioned by Tilburg University (where Stapel was employed when the fraud was uncovered) and encompassed investigators from Tilburg as well as Stapel's previous places of employment, the universities of Amsterdam and Groningen. Levelt, Noort and Drenth Committees, *Flawed Science: The Fraudulent Research Practices of Social Psychologist Diederik Stapel* (University of Tilburg, 2012), https://pure.mpg.de/rest/items/item_1569964/component/file_1569966/content.

7. *The Colbert Report*, January 27, 2011, Comedy Central.

8. See Daryl J. Bem, "Feeling the Future: Experimental Evidence for Anomalous Retroactive Influences on Cognition and Affect," *Journal of Personality and Social Psychology* 100, no. 3 (2011): 407–425, https://doi.org/10.1037/a0021524.

9. Ibid., 410.

10. CBS poll, "Poll: Most Believe in Psychic Phenomena," accessed January 27, 2022, https://www.cbsnews.com/news/poll-most-believe-in-psychic-phenomena/. Other surveys are cited in Sander van der Linden, "How Come Some People Believe in the Paranormal?" *Scientific American*, September 1, 2015, https://www.scientificamerican.com/article/how-come-some-people-believe-in-the-paranormal/.

11. Joseph Jastrow, *Fact and Fable in Psychology* (New York: Houghton Mifflin, 1901), 55.

12. Ibid, 74.

13. Eric-Jan Wagenmakers, Ruud Wetzels, Denny Borsboom, and Han L. J. van der Maas, "Why Psychologists Must Change the Way They Analyze Their Data: The Case of Psi: Comment on Bem (2011)," *Journal of Personality and Social Psychology* 100, no. 3 (2011): 426–432, https://doi.org/10.1037/a0022790.

14. For additional discussion of statistical reasons to question Bem's findings, see Gregory Francis, "Too Good to Be True: Publication Bias in Two Prominent Studies from Experimental Psychology," *Psychonomic Bulletin & Review* 19, no. 2 (2012): 151–156, https://doi.org/10.3758/s13423-012-0227-9.

15. Klaus Fiedler and Joachim I. Krueger, "Afterthoughts on Precognition: No Cogent Evidence for Anomalous Influences of Consequent Events on Preceding Cognition," *Theory and Psychology* 23, no. 3 (2013): 323–333, https://doi.org/10.1177/0959354313485504.

16. Fiedler and Krueger, "Afterthoughts on Precognition," 326.

17. Arthur Conan Doyle, *The Sign of Four* (London: Penguin Classics, 2001).

Chapter 9

1. Paul Lodder, How Hwee Ong, Raoul P. P. P. Grasman, and Jelte M. Wicherts, "A Comprehensive Meta-Analysis of Money Priming," *Journal of Experimental Psychology: General* 148, no. 4 (2019): 688–712, https://doi.org/10.1037/xge0000570.

2. This extraordinary research was published in the highly prestigious journal *Science* by Kathleen D. Vohs, Nicole L. Mead, and Miranda R. Goode, "The Psychological Consequences of Money," *Science* 314, no. 5802 (2006): 1154–1156, https://doi.org/10.1126/science.1132491. The study on children's selfishness is reported in Agata Gąsiorowska, Lan Nguyen Chaplin, Tomasz Zaleskiewicz, and Sandra Wygrab, and Kathleen D. Vohs, "Money Cues Increase Agency and Decrease Prosociality among Children: Early Signs of Market-Mode Behaviors," *Psychological Science* 27, no. 3 (2016): 331–344, https://doi.org/10.1177/0956797615620378.

3. Much of this research on the effects of "achievement" primes has been conducted by Gary Latham and his colleagues. See Xiao Chen, Gary P. Latham, Ronald F. Piccolo, and Guy Itzchakov, "An Enumerative Review and a Meta-Analysis of Primed Goal Effects on Organizational Behavior," *Applied Psychology* 70, no. 1 (2020): 216–253, https://doi.org/10.1111/apps.12239. For a critical review of this research, including a demonstration that the studies manifest funnel-plot asymmetry similar to that shown in figure 8.1, see David R. Shanks and Miguel A. Vadillo, "Publication Bias and Low Power in Field Studies on Goal Priming," *Royal Society Open Science* 8 (2021): 210544, https://doi.org/10.1098/rsos.210544.

4. See Eugene M. Caruso, Kathleen D. Vohs, Brittani Baxter, and Adam Waytz, "Mere Exposure to Money Increases Endorsement of Free-Market Systems and Social Inequality," *Journal of Experimental Psychology: General* 142, no. 2 (2013): 301–306, https://doi.org/10.1037/a0029288.

5. John Bargh is probably the most influential proponent of the view that priming effects are both ubiquitous and (often) unconscious. For instance, see John A. Bargh, "The Historical Origins of Priming as the Preparation of Behavioral Responses: Unconscious Carryover and Contextual Influences of Real-World Importance," *Social Cognition* 32 (2014): 209–224, https://doi.org/10.1521/soco.2014.32.supp.209; John A. Bargh, *Before You Know It: The Unconscious Reasons We Do What We Do* (New York: Simon & Schuster, 2017); John A. Bargh and Tanya L. Chartrand, "The Mind in the Middle: A Practical Guide to Priming and Automaticity Research," in *Handbook of Research Methods in Social and Personality Psychology*, 2nd ed., ed. Harry T. Reis and Charles M. Judd (New York: Cambridge University Press, 2014), 311–344. When money priming studies probe participants to see if they are aware of the influence of the prime on their behavior, they typically report no such awareness; for example, see Leonie Reutner, Jochim Hansen, and Rainer Greifeneder, "The Cold Heart: Reminders of Money Cause Feelings of Physical Coldness," *Social Psychological and Personality Science* 6, no. 5 (2015): 490–495, https://doi.org/10.1177/1948550615574005.

6. Doug Rohrer, Harold Pashler, and Christine R. Harris, "Discrepant Data and Improbable Results: An Examination of Vohs, Mead, and Goode (2006)," *Basic and Applied Social Psychology* 41, no. 4 (2019): 263–271, https://doi.org/10.1080/01973533.2019.1624965.

7. The data in this figure are from a meta-analysis by Miguel A. Vadillo, Tom E. Hardwicke, and David R. Shanks, "Selection Bias, Vote Counting, and Money-Priming Effects: A Comment on Rohrer, Pashler, and Harris (2015) and Vohs (2015)," *Journal of Experimental Psychology: General* 145, no. 5 (2016): 655–663, http://dx.doi.org/10.1037/xge0000157.

8. Lodder et al., "A Comprehensive Meta-Analysis."

9. Ibid.

10. The original report of unconscious flag priming is Travis J. Carter, Melissa J. Ferguson, and Ran R. Hassin, "A Single Exposure to the American Flag Shifts Support toward Republicanism up to 8 Months Later," *Psychological Science* 22 (2011): 1011–1018, https://doi.org/10.1177%2F0956797611414726. The later report on the contents of this group's file drawer is Travis J. Carter, Gayarthri Pandey, Niall Bolger, Ran R. Hassin, and Melissa J. Ferguson, "Has the Effect of the American Flag on Political Attitudes Declined over Time? A Case Study of the Historical Context of American Flag Priming," *Social Cognition* 38, no. 6 (2020): 489–520, https://doi.org/10.1521/soco.2020.38.6.489.

11. A large but unsuccessful attempt to replicate flag priming was reported by Richard A. Klein et al., "Investigating Variation in Replicability: A "Many Labs" Replication

Project," *Social Psychology* 45, no. 3 (2014): 142–152, http://dx.doi.org/10.1027/1864-9335/a000178. Klein's study is described in much more detail in the next chapter.

12. Joshua R. Polanin, Emily E. Tanner-Smith, and Emily A. Hennessy, "Estimating the Difference between Published and Unpublished Effect Sizes: A Meta-Review," *Review of Educational Research* 86, no. 1 (2016): 207–236, https://doi.org/10.3102/0034654315582067.

13. Readers interested in finding out more about these priming effects and evidence that they are at best elusive will be able to access the relevant studies from these sources. For romantic priming, see David R. Shanks et al., "Romance, Risk, and Replication: Can Consumer Choices and Risk-Taking Be Primed by Mating Motives?" *Journal of Experimental Psychology: General* 144, no. 6 (2015): E142–E58, http://dx.doi.org/10.1037/xge0000116. For religious priming, see Shoko Watanabe and Sean M. Laurent, "Past Its Prime? A Methodological Overview and Critique of Religious Priming Research in Social Psychology," *Journal for the Cognitive Science of Religion* 6, no. 1–2 (2021): 31–55, https://doi.org/10.1558/jcsr.38411. For intelligence priming, see Michael O'Donnell et al., "Registered Replication Report: Dijksterhuis and van Knippenberg (1998)," *Perspectives on Psychological Science* 13, no. 2 (2018): 268–294, https://doi.org/10.1177/1745691618755704. A meta-analysis of the effects of scents on consumer behavior was conducted by Holger Roschk and Masoumeh Hosseinpour, "Pleasant Ambient Scents: A Meta-Analysis of Customer Responses and Situational Contingencies," *Journal of Marketing* 84, no. 1 (2019): 125–145, https://doi.org/10.1177/0022242919881137. Although this analysis found an aggregate effect of scents on behavior, the authors reported no tests for publication bias and have not made their data set available for further exploration.

14. This term was introduced by Simmons and colleagues in an article that, probably more than any other, shone a spotlight on biases in the research process in psychology: Joseph Simmons, Leif D. Nelson, and Uri Simonsohn, "False-Positive Psychology: Undisclosed Flexibility in Data Collection and Analysis Allows Presenting Anything as Significant," *Psychological Science* 22 (2011): 1359–1366, https://doi.org/10.1177/0956797611417632.

15. A compelling example in the domain of yet another variety of priming (power priming) is provided by Thandiwe S. E. Gilder and Erin A. Heerey, "The Role of Experimenter Belief in Social Priming," *Psychological Science* 29 (2018): 403–417, https://doi.org/doi.org/10.1177/0956797617737128. For further discussion, see Olivier Klein et al., "Low Hopes, High Expectations: Expectancy Effects and the Replicability of Behavioral Experiments," *Perspectives on Psychological Science* 7 (2012): 572–584, https://doi.org/10.1177/1745691612463704.

16. Daryl J. Bem, "Writing the Empirical Journal Article," in *The Compleat Academic: A Career Guide*, 2nd ed., ed. John M. Darley, Mark P. Zanna, and Henry L. Roediger III (Washington, DC: American Psychological Association, 2003), 185–219.

17. To be fair, researchers are now more and more commonly completing standardized checklists when they publish their research. These checklists comprise statements confirming or clarifying basic aspects of the research such as whether all outcomes have been reported, whether the hypotheses and data analyses were preregistered, and so on. These answers (or indeed just the simple statement, "For all experiments reported in this article, we report how we determined our sample size, all data exclusions, all manipulations, and all measures") then provide some reassurance to anyone reading the article and allow a more valid assessment of the credibility of the findings. See Balazs Aczel et al., "A Consensus-Based Transparency Checklist," *Nature Human Behavior* 4 (2020): 4–6. https://doi.org/10.1038/s41562-019-0772-6.

18. The problems with HARKing, and indeed the term itself, were introduced by Norbert L. Kerr, "HARKing: Hypothesizing after the Results Are Known," *Personality and Social Psychology Review* 2, no. 3 (1998): 196–217, https://doi.org/10.1207/s15327957pspr0203_4.

19. This remarkable claim was made in a thoughtful but provocative article by John P. A. Ioannidis, "Why Most Published Research Findings Are False," *PLOS Medicine* 2 (2005): 696–701, https://doi.org/10.1371/journal.pmed.0020124.

20. For discussion of the interpretation of funnel plot asymmetry, see Jonathan A. C. Sterne et al., "Recommendations for Examining and Interpreting Funnel Plot Asymmetry in Meta-Analyses of Randomised Controlled Trials," *British Medical Journal* 343 (2011): d4002, https://doi.org/10.1136/bmj.d4002.

21. Annie Franco, Neil Malhotra, and Gabor Simonovits, "Underreporting in Psychology Experiments: Evidence from a Study Registry," *Social Psychological and Personality Science* 7, no. 1 (2015): 8–12, https://doi.org/10.1177/1948550615598377.

22. R. Silberzahn et al., "Many Analysts, One Data Set: Making Transparent How Variations in Analytic Choices Affect Results," *Advances in Methods and Practices in Psychological Science* 1 (2018): 337–356, https://doi.org/10.1177/2515245917747646.

23. A reporting checklist designed to provide this transparency for multi-analyst studies has been developed by Balazs Aczel et al., "Consensus-Based Guidance for Conducting and Reporting Multi-Analyst Studies," *eLife* 10 (2021): e72185, https://doi.org/10.7554/eLife.72185.

24. Cardeña's review is in Etzel Cardeña, "The Experimental Evidence for Parapsychological Phenomena: A Review," *American Psychologist* 73, no. 5 (2018): 663–677, https://doi.org/10.1037/amp0000236. Another up-to-date meta-analysis of parapsychological research, which goes to considerable lengths to evaluate and correct for publication bias, is Patrizio E. Tressoldi and Lance Storm, "Stage 2 Registered Report: Anomalous Perception in a Ganzfeld Condition—A Meta-Analysis of More than 40 Years Investigation," *F1000Research* 10 (2021): 234, https://doi.org/10.12688/f1000research.51746.1.

Chapter 10

1. See Brian Deer, "How the Case against the MMR Vaccine Was Fixed," *BMJ* 342 (2011): c5347, https://doi.org/10.1136/bmj.c5347.

2. This estimate is from Matthew C. Makel, Jonathan A. Plucker, and Boyd Hegarty, "Replications in Psychology Research: How Often Do They Really Occur?" *Perspectives on Psychological Science* 7, no. 6 (2012): 537–542, https://doi.org/10.1177/174569161 2460688.

3. This ground-breaking project is reported in Richard A. Klein et al., "Investigating Variation in Replicability: A "Many Labs" Replication Project," *Social Psychology* 45, no. 3 (2014): 142–152, http://dx.doi.org/10.1027/1864-9335/a000178. Reassuringly, ten of the thirteen effects chosen by Klein and colleagues replicated strongly. Of the three that did not, two were priming effects: flag priming (discussed in chapter 9) and money priming.

4. Doug Rohrer, Harold Pashler, and Christine R. Harris. "Do Subtle Reminders of Money Change People's Political Views?" *Journal of Experimental Psychology: General* 144, no. 4 (2015): e73-e85, https://doi.org/10.1037/xge0000058.

5. For a comprehensive review of the history and importance of preregistration as a tool to reduce bias, see Tom E. Hardwicke and Eric-Jan Wagenmakers, "Reducing Bias, Increasing Transparency, and Calibrating Confidence with Preregistration," *Nature Human Behaviour* (2023), https://osf.io/d7bcu/download.

6. For instance, see John K. Sakaluk, "Exploring Small, Confirming Big: An Alternative System to the New Statistics for Advancing Cumulative and Replicable Psychological Research," *Journal of Experimental Social Psychology* 66 (2016): 47–54, https://doi.org/10.1016/j.jesp.2015.09.013.

7. This evidence comes from Robert M. Kaplan and Veronica L. Irvin, "Likelihood of Null Effects of Large NHLBI Clinical Trials Has Increased over Time," *PLOS ONE* 10, no. 8 (2015): e0132382, https://doi.org/10.1371/journal.pone.0132382.

8. See Christopher W. Jones, Lukas G. Keil, Wesley C. Holland, Melissa C. Caughey, and Timothy F. Platts-Mills, "Comparison of Registered and Published Outcomes in Randomized Controlled Trials: A Systematic Review," *BMC Medicine* 13 (2015): 282, https://doi.org/10.1186/s12916-015-0520-3.

9. Sylvain Mathieu, An-Wen Chan, and Philippe Ravaud, "Use of Trial Register Information during the Peer Review Process," *PLOS ONE* 8, no. 4 (2013): e59910, https://doi.org/10.1371/journal.pone.0059910.

10. The ease and impact of retrospective comparisons of preregistrations against published results should not be underestimated and may, for reputational reasons, encourage researchers in the future to be increasingly careful to minimize significant

discrepancies. For example, see Ben Goldacre et al., "COMPare: A Prospective Cohort Study Correcting and Monitoring 58 Misreported Trials in Real Time," *Trials* 20 (2019): 118, https://doi.org/10.1186/s13063-019-3173-2.

11. A review of the history and current status of registered reports, by one of the pioneers of this format, is Christopher D. Chambers and Loukia Tzavella, "The Past, Present and Future of Registered Reports," *Nature Human Behaviour* 6, (2022): 29–42, https://doi.org/10.1038/s41562-021-01193-7.

12. This project is reported in Courtney K. Soderberg et al., "Initial Evidence of Research Quality of Registered Reports Compared with the Standard Publishing Model," *Nature Human Behaviour* 5, no. 8 (2021): 990–997, https://doi.org/10.1038/s41562-021-01142-4. It must be noted that in a postassessment survey, the majority of reviewers acknowledged that they were able to tell the difference between the registered reports and comparison articles, and so were not blind. However registered reports were rated stronger than comparison articles even in those cases where the reviewer did not correctly identify which was which.

13. See Christopher Allen and David M. A. Mehler, "Open Science Challenges, Benefits and Tips in Early Career and Beyond," *PLOS Biology* 17, no. 12 (2019): e3000246, https://doi.org/10.1371/journal.pbio.3000246; Anne M. Scheel, Mitchell R. M. J. Schijen, and Daniël Lakens, "An Excess of Positive Results: Comparing the Standard Psychology Literature with Registered Reports," *Advances in Methods and Practices in Psychological Science* 4, no. 2 (2021), https://doi.org/10.1177/25152459211007467.

14. Allen and Mehler, "Open Science Challenges"; Scheel et al., "An Excess of Positive Results."

15. These are calculated from data compiled by Paul Lodder, How Hwee Ong, Raoul P. P. P. Grasman, and Jelte M. Wicherts, "A Comprehensive Meta-Analysis of Money Priming," *Journal of Experimental Psychology: General* 148, no. 4 (2019): 688–712, https://doi.org/10.1037/xge0000570.

16. This decline effect was first reported by Tom J. Johnsen and Oddgeir Friborg, "The Effects of Cognitive Behavioral Therapy as an Anti-Depressive Treatment Is Falling: A Meta-Analysis," *Psychological Bulletin* 141, no. 4 (2015): 747–768, https://doi.org/10.1037/bul0000015, although their evidence has not gone unquestioned: Ioana A. Cristea, Simona Stefan, Eirini Karyotaki. Daniel David, Seven D. Hollon, and Pim Cuijpers, "The Effects of Cognitive Behavioral Therapy Are Not Systematically Falling: A Revision of Johnsen and Friborg (2015)," *Psychological Bulletin* 143, no. 3 (2017): 326–340, https://doi.org/10.1037/bul0000062.

17. Tom Stafford has made a powerful argument that our experimental practices strikingly distort the apparent importance of different influences on our behavior and make the nonreporting of some influences seem more meaningful than it truly is. Imagine that priming effects are small but real. In an experiment, we administer

a prime to one group of participants but not to another. Because of randomization, our experiment controls all other influences on behavior and isolates the impact of the prime, which (with large enough samples) might be statistically significant. But this effect is at the level of the group: when we consider any given individual, the prime might be much less influential on their behavior than numerous other factors. Hence, nonreporting of this influence, which might seem to indicate unawareness, reveals nothing more than relative unimportance to the individual. The following analogy emphasizes the issue. Imagine that a large medical trial establishes that eating an extra apple a day lowers blood pressure to a statistically significant degree. But if we take an individual at random, it is obvious that compared to their salt intake, exercise, and whether they smoke, eating or not eating an apple is trivial for explaining their blood pressure. Stafford puts it as follows: "The logic of many of our behavioral experiments encourages a perspectival shift in which factors which have the major influence on each individual's choices are rendered invisible, while an experimental factor which has a minor influence on each individual's choice is highlighted" (p. 2). See Tom Stafford, "The Perspectival Shift: How Experiments on Unconscious Processing Don't Justify the Claims Made for Them," *Frontiers in Psychology* 5 (2014): 1067, https://doi.org/10.3389/fpsyg.2014.01067.

18. The findings in this paragraph are based on Tom E. Hardwicke et al., "Citation Patterns Following a Strongly Contradictory Replication Result: Four Case Studies from Psychology," *Advances in Methods and Practices in Psychological Science* 4, no. 3 (2021), https://doi.org/10.1177/25152459211040837.

19. For the evidence and quotations regarding the experiment, see Richard A. Griggs and George I. Whitehead, "Coverage of the Stanford Prison Experiment in Introductory Social Psychology Textbooks," *Teaching of Psychology* 41, no. 4 (2014): 318–324, https://doi.org/10.1177/0098628314549703.

20. See Stephen Reicher and S. Alexander Haslam, "Rethinking the Psychology of Tyranny: The BBC Prison Study," *British Journal of Social Psychology* 45, no. 1 (2006): 1–40, https://doi.org/10.1348/014466605X48998.

21. Jared M. Bartels and Patricia Schoenrade, "The Implicit Association Test in Introductory Psychology Textbooks: Blind Spot for Controversy," *Psychology Learning & Teaching* (2021), https://doi.org/10.1177/14757257211055200.

22. As documented by Luis Morís Fernández, Tom E. Hardwicke, and Miguel A. Vadillo, "Retracted Papers Clinging on to Life: An Observational Study of Post-Retraction Citations in Psychology," PsyArXiv, April 26, 2022, https://psyarxiv.com/cszpy/.

23. The maxim is named after the British economist Charles Goodhart. For evidence that quantity of publications has suffered this fate, see Michael Fire and Carlos Guestrin, "Over-Optimization of Academic Publishing Metrics: Observing Goodhart's Law in Action," *GigaScience* 8, no. 6 (2019), https://doi.org/10.1093/gigascience/giz053.

24. This effect size estimate is based on Fredrik Santoft, Erland Axelsson, Lars-Göran Öst, Maria Hedman-Lagerlöf, Jens Fust, and Erik Hedman-Lagerlöf, "Cognitive Behavior Therapy for Depression in Primary Care: Systematic Review and Meta-Analysis," *Psychological Medicine* 49, no. 8 (2019): 1266–1274, https://doi.org/10.1017/S0033291718004208.

25. R. Chris Fraley and Simine Vazire, "The N-Pact Factor: Evaluating the Quality of Empirical Journals with Respect to Sample Size and Statistical Power," *PLOS ONE* 9, no. 10 (2014): e109019, https://doi.org/10.1371/journal.pone.0109019.

26. Samantha F. Anderson, Ken Kelley, and Scott E. Maxwell, "Sample-Size Planning for More Accurate Statistical Power: A Method Adjusting Sample Effect Sizes for Publication Bias and Uncertainty," *Psychological Science* 28, no. 11 (2017): 1547–1562, https://doi.org/10.1177/0956797617723724, provide an estimated median sample size of about 48 in between-groups experiments in psychological science.

27. On average effect sizes, see T. D. Stanley, Evan C. Carter, and Hristos Doucouliagos, "What Meta-Analyses Reveal about the Replicability of Psychological Research," *Psychological Bulletin* 144, no. 12 (2018): 1325–46. https://doi.org/10.1037/bul0000169. On the (lack of) change in statistical power over the past half-century, see Paul E. Smaldino and Richard McElreath, "The Natural Selection of Bad Science," *Royal Society Open Science* 3 (2016): 160384, https://doi.org/10.1098/rsos.160384.

28. See Andreas Lundh, Joel Lexchin, Barbara Mintzes, Jeppe B. Schroll, and Lisa Bero, "Industry Sponsorship and Research Outcome," *Cochrane Database of Systematic Reviews* (2017): MR000033, https://doi.org/10.1002/14651858.MR000033.pub3.

29. Examples include: Ap Dijksterhuis, Maarten W. Bos, Loran F. Nordgren, and Rick B. van Baaren, "On Making the Right Choice: The Deliberation-without-Attention Effect," *Science* 311, no. 5763 (2006): 1005–1007, https://doi.org/10.1126/science.1121629, with unsuccessful replication by Mark R. Nieuwenstein et al., "On Making the Right Choice: A Meta-Analysis and Large-Scale Replication Attempt of the Unconscious Thought Advantage," *Judgment and Decision Making* 10, no. 1 (2015): 1–17; Lawrence E. Williams and John A. Bargh, "Experiencing Physical Warmth Promotes Interpersonal Warmth," *Science* 322, no. 5901 (2008): 606–607, https://doi.org/10.1126/science.1162548, with unsuccessful replication by Christopher F. Chabris, Patrick R. Heck, Jaclyn Mandart, Daniel J. Benjamin, and Daniel J. Simons, "No Evidence That Experiencing Physical Warmth Promotes Interpersonal Warmth: Two Failures to Replicate Williams and Bargh (2008)," *Social Psychology* 50, no. 2 (2019): 127–132, https://doi.org/10.1027/1864-9335/a000361; and Antoine Bechara, Hanna Damasio, Daniel Tranel, and Antonio R. Damasio, "Deciding Advantageously before Knowing the Advantageous Strategy," *Science* 275, no. 5304 (1997): 1293–1295, https://doi.org/10.1126/science.275.5304.1293, with unsuccessful replication by Tiago V. Maia and James L. McClelland, "A Reexamination of the Evidence for the Somatic Marker Hypothesis: What Participants Really Know in the Iowa Gambling Task," *Proceedings of the National Academy of Sciences* 102, no. 45 (2004): 16075–16080, https://doi.org/10.1073/pnas.0406666101.

30. Fraley and Vazire, "The N-Pact Factor."

31. It must be acknowledged that everything else is rarely equal and that a poorly designed study with 10,000 participants is still a poorly designed study. Also larger sample sizes do not necessarily entail greater statistical power, which depends on the exact experimental design. For instance, within-subjects studies in many branches of cognitive psychology can get away with small sample sizes by averaging across many measurements from each experimental participant. For a thoughtful discussion, see Philip L. Smith and Daniel R. Little, "Small Is Beautiful: In Defense of the Small-N Design," *Psychonomic Bulletin & Review* 25, 2083–2101 (2018), https://doi.org/10.3758/s13423-018-1451-8.

32. For coercive citation, see Allen W. Wilhite and Eric A. Fong, "Coercive Citation in Academic Publishing," *Science* 335, no. 6068 (2012): 542–543, https://doi.org/10.1126/science.1212540. For journal reluctance to publish corrections or retractions, see Goldacre et al., "COMPare."

33. Johannes Bohannon, Diana Koch, Peter Homm, and Alexander Driehaus, "Chocolate with High Cocoa Content as a Weight-Loss Accelerator," *International Archives of Medicine* 8 (2015).

34. For instance, compelling evidence for this from the field of economics is reported by Nicholas Swanson et al., "Research Transparency Is on the Rise in Economics," *AEA Papers and Proceedings* 110 (2020): 61–65, https://doi.org/10.1257/pandp.20201077.

Chapter 11

1. "World Health Organisation—COVID-19-China," accessed April 29, 2022, https://www.who.int/emergencies/disease-outbreak-news/item/2020-DON229.

2. "Johns Hopkins University and Medicine Coronavirus Resource Centre," accessed March 31, 2022, https://coronavirus.jhu.edu/map.html.

3. The discussion of the application of TPB to social distancing is based on data reported in Laurel P. Gibson, Renee E. Magnan, Emily B. Kramer, and Angela D. Bryan, "Theory of Planned Behavior Analysis of Social Distancing during the Covid-19 Pandemic: Focusing on the Intention–Behavior Gap," *Annals of Behavioral Medicine* 55, no. 8 (2021): 805–812, https://doi.org/10.1093/abm/kaab041.

4. Liat Ayalon et al., "A Systematic Review of Existing Ageism Scales," *Ageing Research Reviews*, 54 (2019): 100919, https://doi.org/10.1016/j.arr.2019.100919.

5. This example is discussed at greater length in Markus I. Eronen and Laura F. Bringmann, "The Theory Crisis in Psychology: How to Move Forward," *Perspectives on Psychological Science* 16, no. 4 (2021): 779–788, https://doi.org/10.1177/1745691620970586.

6. The very large field within the behavioral sciences that is devoted to designing and validating tests is called *psychometrics*. Decades of work in psychometrics

have shown that good tests—ones that validly measure the underlying construct (such as an unconscious attitude) that they seek to measure—require a number of features. These are best described as distinct varieties of validity. A good test must predict meaningful aspects of behavior and yield scores that are similar to other tests designed to measure the same construct (convergent validity) while differing from tests that measure other constructs (discriminant validity). If the IAT provides a good measure of unconscious attitudes, then it should generate scores across a sample of individuals that, on the one hand, correlate with scores from other tests designed to measure these same unconscious attitudes, while on the other hand do not correlate with scores from tests designed to measure other constructs (conscious attitudes being the obvious case here). At the same time, IAT scores should predict some aspect of behavior such as job applicant ratings (predictive validity). Once it has been established that a test is psychometrically sound, it can be administered to a large sample of individuals, alongside other tests, and the results used to build a structural equation model or *nomological net* of the sort derived from the COVID-19 social distancing example (figure 11.1).

7. Ulrich Schimmack, "The Implicit Association Test: A Method in Search of a Construct," *Perspectives on Psychological Science* 16 (2021): 396–414, https://doi.org/10.1177/1745691619863798; Edouard Machery, "Anomalies in Implicit Attitudes Research," *WIREs Cognitive Science* 13, no. 1 (2022): e1569, https://doi.org/10.1002/wcs.1569.

8. The project is described in Justin F. Landy et al., "Crowdsourcing Hypothesis Tests: Making Transparent How Design Choices Shape Research Results," *Psychological Bulletin* 146 (2020): 451–479, https://doi.org/10.1037/bul0000220.

9. Ulrich Schimmack, "The Validation Crisis in Psychology," *Meta-Psychology* 5 (2021), https://doi.org/10.15626/MP.2019.1645.

10. Guillermo Campitelli and Paul Gerrans, "Does the Cognitive Reflection Test Measure Cognitive Reflection? A Mathematical Modeling Approach," *Memory & Cognition* 42, no. 3 (2014): 434–447, https://doi.org/10.3758/s13421-013-0367-9; Caterina Primi, Kinga Morsanyi, Francesca Chiesi, Maria Anna Donati, and Jayne Hamilton, "The Development and Testing of a New Version of the Cognitive Reflection Test Applying Item Response Theory (IRT)," *Journal of Behavioral Decision Making* 29, no. 5 (2016): 453–469, https://doi.org/10.1002/bdm.1883.

11. The illustration of these concepts using the fictional embodiment priming theory is further developed in Klaus Oberauer and Stephan Lewandowsky, "Addressing the Theory Crisis in Psychology," *Psychonomic Bulletin & Review* 26, no. 5 (2019): 1596–1618, https://doi.org/10.3758/s13423-019-01645-2. Additional discussion on the challenges of theory building can be found in Iris van Rooij and Giosuè Baggio, "Theory before the Test: How to Build High-Verisimilitude Explanatory Theories in Psychological Science," *Perspectives on Psychological Science* 16, no. 4 (2021): 682–697, https://doi.org/10.1177/1745691620970604.

12. Oberauer and Lewandowsky, "Addressing the Theory Crisis." For a forceful critique of preregistration see Aba Szollosi, et al., "Is Preregistration Worthwhile?" *Trends in Cognitive Sciences* 24, no. 2 (2020): 94–95, https://doi.org/10.1016/j.tics.2019.11.009.

13. Ivan Grahek, Mark Schaller, and Jennifer L. Tackett, "Anatomy of a Psychological Theory: Integrating Construct-Validation and Computational-Modeling Methods to Advance Theorizing," *Perspectives on Psychological Science* 16, no. 4 (2021): 803–815, https://doi.org/10.1177/1745691620966794.

14. For instance, in the domain of memory and amnesia, a model with a single latent and conscious memory process has been shown to be sufficient to explain complex patterns of data: Christopher J. Berry, David R. Shanks, Maarten Speekenbrink, and Richard N. A. Henson, "Models of Recognition, Repetition Priming, and Fluency: Exploring a New Framework," *Psychological Review* 119, no. 1 (2012): 40–79, https://doi.org/10.1037/a0025464.

15. Eronen and Bringmann, "The Theory Crisis."

16. "1918 Pandemic (H1N1 virus)," accessed April 29, 2022, https://www.cdc.gov/flu/pandemic-resources/1918-pandemic-h1n1.html.

17. George A. Soper, "The Lessons of the Pandemic," *Science* 49, no. 1274 (1919): 501–506, https://www.jstor.org/stable/1642775.

18. Soper, "The Lessons."

19. For instance, Jay J. Van Bavel et al., "Using Social and Behavioural Science to Support COVID-19 Pandemic Response," *Nature Human Behaviour* 4 (2020): 460–471, https://doi.org/10.1038/s41562-020-0884-z; Robert West, Susan Michie, G. James Rubin, and Richard Amlôt, "Applying Principles of Behavior Change to Reduce SARS-CoV-2 Transmission," *Nature Human Behavior* 4 (2020): 451–459, https://doi.org/10.1038/s41562-020-0887-9.

Index

Note: *Italic* page numbers denote figures and their captions.